The Use of Information and Communication Technologies (ICT) in the Management of the Innovative and Smart City

Key elements in the formation of smart cities are development based on cooperation between local governments and business, the realization of sustainable development goals, the creation of social sustainability, the advancement of information and communication technologies (ICT) and the use of innovative solutions. The theories of information and communication have become the main driver of economic development in a globalized world and the information component plays a key role in building the competitive potential of countries and cities.

The purpose of this book is to systematize knowledge about the practical application and importance of innovative solutions in the economic and social fields. An important aspect of this edition is the presentation of innovations applied in the field of conservation, not only of energy or environmental resources, but also of economic, social and spatial resources. We propose that this edition provides a knowledge base on the basic concepts of corporate social responsibility (CSR), formulating the important role of management in the context of CSR, smart management, smart economy and smart cities.

The book is a comprehensive study of the smart city concept in its broadest sense, covering environmental, social and economic issues.

It will be helpful to professionals in companies, in logistics departments and in the management teams. It can be useful for practitioners and analysts in the transportation industry to expand applications of ICT solutions. It will also be valuable to researchers, doctoral students of economics in management and entrepreneurship, logistics, as well as students at the advanced level, as it brings a range of valuable information that can serve as base material and reference for further research, programmes and studies.

The Use of Information and Communication Technologies (ICT) in the Management of the Innovative and Smart City

Edited by
Judyta Kabus, Luiza Piersiala,
and Michał Dziadkiewicz

CRC Press
Taylor & Francis Group
Boca Raton London New York

CRC Press is an imprint of the
Taylor & Francis Group, an **informa** business

Designed cover image: © Shutterstock

First edition published 2025
by CRC Press
2385 NW Executive Center Drive, Suite 320, Boca Raton FL 33431

and by CRC Press
4 Park Square, Milton Park, Abingdon, Oxon, OX14 4RN

CRC Press is an imprint of Taylor & Francis Group, LLC

ISBN: 978-1-032-73628-0 (hbk)
ISBN: 978-1-032-73630-3 (pbk)
ISBN: 978-1-003-46515-7 (ebk)

DOI: 10.1201/9781003465157

Typeset in Adobe Caslon Pro
by KnowledgeWorks Global Ltd.

Contents

Preface

Dear Reader,

We invite you to read a book that will take you into the world of innovation, sustainability and modern technology in the context of urban life. The publication focuses on the dynamics of modern cities, where sustainability, smart solutions and social responsibility are shaping new standards of urban life.

Cities are currently facing many challenges, including climate change and technological development. Smart cities are not an end in themselves – they are just a slogan, an idea that brings together many activities leading to the development of cities where life will be more pleasant, healthy, comfortable and cheaper. To achieve this goal, many paths must be followed simultaneously. The smart city concept encompasses various models related to service innovation, urban maturity and public participation.

The book is an attempt to systematize knowledge about the practical application and importance of innovative solutions in the economic and social fields. An important aspect of this edition is the presentation of innovations applied in the field of conservation, not only of energy or environmental resources, but also of economic, social and spatial resources. We propose that this issue provides a knowledge base on the basic concepts of corporate social responsibility (CSR), formulating the important role of governance in the context of CSR, smart governance, smart economy and smart cities.

We present a scientific discussion covering the following topics: the use of information and communication technologies (ICT) in the management of the transport, shipping and logistics (TSL) industry; innovative and smart cities of the future; urban laboratories as a tool for improving the quality of life of city residents in accordance with the smart city concept; the use of digital technologies and open public data to improve decision-making processes and living standards; the role of corporate governance in CSR; and solutions to support zero-carbon energy generation for society.

We believe that the book will be helpful to professionals in companies, logistics departments, and management teams. It can be useful for practitioners and analysts in the transportation industry to expand applications of ICT solutions. It will also be valuable to researchers, doctoral students of economics in management and entrepreneurship, logistics, as well as students at the advanced level, as it brings a wealth of valuable information that can serve as base material and reference for further research, programmes and studies.

Let this book become a guide for you along the paths of urban evolution, and let its contents open up new horizons for thinking about the future of our urban communities.

About the Editors

Judyta Kabus is an Assistant Professor at the Częstochowa University of Technology. She has doctorate degrees in Humanities and in Management and Quality Sciences. She is the President of the Association of Women of the Częstochowa Region.

She has authored and co-authored more than 100 scientific publications. She is the co-author of two books in the area of logistics and management and the editor of multi-author monographs. She is the technical editor of Scientific Journals of the Częstochowa University of Technology, and reviewer of scientific works in Polish and foreign journals. She is also the author of teaching materials for learning German and education for the development of research. Dr. Kabus is responsible for organizing national and international conferences and she is winner of the Rector's Awards of the Częstochowa University of Technology. She has been a long-term scientific collaborator with the Technical University of Ostrava and participant of the Erasmus+ student exchange program. She is a member of the Organisation and Management Scientific Society (TNOiK) and bilateral research projects. She also has been field-specific supervisor of student internships and member of several projects. She is the Local Project Coordinator of the Model location, implementation of the Częstochowa City project "Engaging stakeholders for knowledge-based, sustainable and socially responsible management of urban housing."

Dr. Kabus is manager of the Excellent Science II project funded by the Ministry of Higher Education and Science and the coordinator of the research team at the Faculty of Management, Częstochowa University of Technology. She is also scientific and industrial expert at the National Center for Research and Development. Her scope of work includes innovation, social responsibility, sustainable development, green economy, market trends and urban logistics.

Luiza Piersiala is Assistant Professor at the Częstochowa University of Technology. She has doctorate degree in Management and Quality Sciences, and is member of the Association of Women of the Częstochowa Region.

She is the author and co-author of more than 50 scientific publications. She has also co-authored a script for students in the area of logistics and accounting and edited multi-author monographs. Dr. Piersiala is a reviewer of scientific works in Polish and foreign journals. She is also a participant of a foreign internship at the Technical University of Ostrava. She has organized a national conference and has won multiple Rector's Awards of the Częstochowa University of Technology. She is supervisor of the "Lean&Smart" Student Scientific Club.

Her scope of work includes entrepreneurship, innovation, social responsibility, sustainable development, special economic zones and consumer behavior.

Michał Dziadkiewicz is an Assistant Professor at the Częstochowa University of Technology. He has a doctorate degree in Economic Sciences, and is attorney-at-law specializing in matters of civil law, covering, in particular, the issues of real property management, including the real property of Housing Communities, as well as the real property comprising the commune's housing stock, debt collection. He conducts scientific research on, inter alia, aspects of management of commune housing resources, rent collection, prevention of social exclusion, as well as sports law and legal conditions of physical recreation. He is the author of several dozen publications, a reviewer of scientific papers in Polish and foreign journals, and an organizer of national and international conferences. He completed postgraduate studies in economics and management, specializing in managerial economics and management.

He is an expert in the Organisation and Management Scientific Society (TNOiK). His scope of work includes management of commune housing resources, innovation, social responsibility, sustainable development and smart cities.

Contributors

Izabela Bagińska
Department of Economics and Finance
Faculty of Law and Economics
Jan Dlugosz University
Częstochowa, Poland

Beata Barszczowska
Katowice Business University
Katowice, Poland

Jana Chovancová
Faculty of Management, Economics and Business
University of Presov
Presov, Slovakia

Marcin Dziadkiewicz
Faculty of Law and Administration
University of Warsaw
Warszawa, Poland

Judyta Kabus
Department of Logistics
Faculty of Management
Częstochowa University of Technology
Częstochowa, Poland

Paulina Kaleja
Department of Advanced Energy Technologies
Faculty of Infrastructure and Environment
Częstochowa University of Technology
Częstochowa, Poland

Patrycja Krawczyk
Department of Finance, Banking and Accounting
Faculty of Management
Częstochowa University of Technology
Częstochowa, Poland

Luiza Piersiala
Department of Logistics
Faculty of Management
Częstochowa University of Technology
Częstochowa, Poland

Janusz Stępnik
Department of Advanced Energy Technologies
Faculty of Infrastructure and Environment
Częstochowa University of Technology
Częstochowa, Poland

Agnieszka Ulfik
Department of Management Theory
Faculty of Sport and Tourism Management
Academy of Physical Education in Katowice
Katowice, Poland

Renata Włodarczyk
Department of Advanced Energy Technologies
Faculty of Infrastructure and Environment
Częstochowa University of Technology
Częstochowa, Poland

1

Information and Communication Technology (ICT) Challenges for the TSL Industry

Polish Example

PATRYCJA KRAWCZYK

1.1 Introduction

In this global world, information and communication theories have become major drivers of economic development. In building the competitive potential of countries and developing international relations, the information component plays a key role (Petkova, Ryabokon and Vdovychenko, 2019). The rapid worldwide progress of information and communication technologies (ICT) in the past three decades has attracted increasing attention among economists and researchers who have focused on studying the impact of ICT diffusion on the enterprises (Bahrini and Qaffas, 2019). Subsequent years bring new developments in this area. The challenge is their proper implementation.

After the fall of communism in 1989, Poland became one of the best performing countries in the group of post-communist economies. Following the period of political change, enterprise restructuring and changes in the country's economic policy, especially after Poland's accession to the European Union (EU), the country became an attractive investment location mainly because of low costs of labour, a wide market for products and services, and an increased internal demand (Piersiala, 2018). Arendt and Garbowski devoted their study to the role of ICT in explaining the innovativeness of Polish enterprises (Arendt and Grabowski, 2019).

The economic environment comprises many crises, and rapid changes in competitors' conditions create a very dynamic and unstable

DOI: 10.1201/9781003465157-1

view of the reality where the strategies of the future development are realized. This unstable, changeable state creates the frames of transport functioning that operate the transport needs of the society. A vigorous transport industry determines the further economic development in Poland. The transport industry has a very big contribution in the branch structure of the Polish economy and its role is still growing. Among the main factors that affect the evolution of transport is the development of consumerism, globalization, as well as the modernization of road infrastructure (Krawczyk and Kokot-Stępień, 2020).

There are many benefits to be gained from the use of ICT systems. A fundamental one is achieving business goals more efficiently, finding systemic solutions in operations and responding to various crises in a more effective manner (Osinska and Zalewski, 2020). Unfortunately, compared to other European countries, Polish companies seem to lag behind when it comes to the use of ICT.

These conclusions lead to the main question: What is the level of usage of ICT among transport, shipping and logistics (TSL) enterprises in Poland? The presented issue shall be verified through the following detailed research inquiries:

1. What is the significance of the TSL sector for the Polish economy?
2. What is the level of use of ICT in Poland compared to other countries?
3. What challenges the TSL industry faces in ICT usage?

The following issues related to ICT will be analysed in detail: access to the Internet in enterprises, employees with Internet access, website, advertising on the Internet, social media, public open data, data utilization, electronic sales, cloud computing, artificial intelligence and digital intensity index.

The analysis shall be based upon a critical analysis of literature, branch-specific reports as well as observations of practice, own experience and thoughts.

1.2 Literature Review

ICT is a tool used in many areas, including scientific research. Most scientists treat ICT as basic equipment and use them without realizing their significance for their work. However, many have discussed the topic of the significance of ICT from the point of view of their field of research, e.g. engineering, medicine or economy. Regarding the growing importance of ICT and the way it is transforming the world, many academicians and researchers have focused on studying the impact of ICT on economic growth at the industry level, at the national level and at the cross-country level (Bahrini and Qaffas, 2019).

Taylor and Ciechanski write about the transformation of Polish transport companies after the year 1989. Their paper seeks to reconstruct post-1989 organizational and ownership transformations in Poland's rail-, road- and urban-transport companies, as well as those involved in inland shipping (Ciechański and Taylor, 2017). Some papers present the most important factors influencing contemporary changes in transport needs, such as the progress of globalization and integration, the development of e-commerce and logistics (Ciechański, 2020). We can also see that there is a synthetic statistical profile of transport development in Poland after 2005. They also consider the challenges for the transport industry resulting from globalization. They emphasize on liberalization and globalization that consequently lead to development. The diagnosis of the Polish TSL sector announces that it does not fully use its potential. Its development after the year 1989 was dynamic, but now it faces new challenges (Krawczyk, 2018).

Transportation plays a vital role in shaping economies. Transportation supports clusters and agglomerations, increases productivity, enhances jobs and labour market accessibility, pens new markets for businesses and enhances supply chain efficiency (Comtois, Rodrigue and Slack, 2016). The concept of transport is very wide and can be considered with many areas of the economy. Along with the systemic changes and economic transformations, the transport policy has also changed.

Today, the world is developing in a multidimensional fashion. A highly developed transport infrastructure creates a competitive edge. Transport is the social/economic activity which determines any and all other activities of people and businesses. Transport is characterized by an organizational/legal complexity, diversity of processes relating to movement of cargo and people as well as a complexity of technical correlations. Transport enables the economic development of a country, its foreign trading and its market integration; it ensures accessibility to different regions. Transport affects economic development, division of work, specialization of production, and it increases the reach of the market. It is decisive, in terms of success, of not only one economic sector but the economic success of the entire country. Transport significantly affects the gross domestic product (GDP) and the development of various national economy sectors (Button, 2010). Transport affects sustainable development, which refers to social, ecological, economic, spatial and functional factors; and the growth and effective functioning of the transport system greatly facilitates conducting business operations.

The transportation and logistics sector is going through a challenging time of digitization. New technical devices are being developed to help solve emerging problems. ICT is expected to help properly prepare for future business trends (Burinskiene, 2023).

Contemporary economic processes have been accompanied by a significant increase in mobility and higher levels of accessibility. Globalization and international integration of economies create favourable conditions for more dynamic development. Polish economy can be considered as an interesting case study for the proposed scientific problem, as it is the biggest economy in Central Europe.

It has become common to forecast demand for transport of goods due to the increase in the main macroeconomic indicators, in particular GDP. It is done with the assumption that there is a balanced proportion between the intensity of physical production and exchange and the intensity of demand for transport. It is still worth to confront this point of view with the statistical data from the previous periods. This kind of data show that in many cases the correlation with macroeconomic values and transport is very weak or even negative. The volume of demand for transport is affected not only by the level of physical production and exchange, but also by changes in business, location, optimization of transport and logistics processes, changes in technology and production organization, inventory management and other quantitative and qualitative factors (Burnewicz, 2017). Czech and Lewczuk explore taxonomic and econometric analysis of road transport development in Poland. The results obtained in this research show that there is an interdependence in the area of socioeconomic development, in particular, voivodships and the level of their road development. The higher the economic parameters of the related position of a voivodship in the ranking, the higher is the level of road transport development (Czech, Bołtromiuk and Lewczuk, 2016).

An effective tool used for collecting, storing, processing, sending and presenting information is ICT. Unfortunately, entrepreneurs sometimes perceive implementing modern IT solutions as a cost and do not notice that they can be an effective tool to facilitate business management business. Most prominent barriers in the implementation of new technologies are high costs of necessary infrastructure and some applications as well as data security concerns (Petryczka, 2017).

Artificial intelligence (AI) technologies are on the rise in the field of transport systems around the world (Abduljabbar et al., 2019; Mclean et al., 2023). AI is changing the way we think about TSL sector. The desire to improve the safety, efficiency and sustainability of road transport is the main motivation behind the rapid growth in the adoption of AI technologies (Furlan et al., 2020; Eskandari Torbaghan et al., 2022). Applying AI to transport sector brings many benefits. It optimizes logistics processes, reduces repetitive activities, collects and processes extensive data (e.g. on costs and delivery times) and helps manage risk. We can read about the potential applications of AI, such as AI-driven chatbots for education and research in air transport management, with a use case of the recently published ChatGPT (Sun, Wandelt and Zhang, 2023).

1.3 Diagnosis of the TSL Brand Situation in Poland

Transport in the EU contributes around 5% of GDP and employs around 10 million people. It is crucial to the functioning of European businesses and global supply chains (Olejniczak and Dębicka, 2021). Poland's accession to the EU brought many changes in the functioning of the Polish TSL sector. Although some attempts to liberalize the market had been made much earlier, it was EU accession that played a decisive role in the changes. Thanks to the opening of the borders and the abolition of customs duties, the progress of the TSL industry in Poland became more harmonious and visible.

The TSL sector is an important part of the Polish economy and the share of transport and logistics (goods and people) in Poland's GDP is estimated to be around 11.5% (Fechner and Szyszka, 2018). The Polish TSL sector dominates the European market. It has even overtaken companies from Germany. Transportation and warehousing remains one of the most important parts of the Polish economy, generating about 6% of Polish GDP and employing about 6% of the Polish workforce. Of the total workforce in this sector 51% is engaged in road transport of goods, 16% in warehousing, and 11% in courier and postal services – the most important components of the TSL industry. Poland's transport sector is based on 165,000 micro-enterprises, employing 303,000 people. Only in two countries (Spain and Greece) is the share of micro-companies in this market larger.

Logistics Performance Index (LPI) ranking is published since 2007 by the World Bank, and it assesses a given country in six areas namely: efficiency of the customs clearance process, quality of infrastructure, ease of organizing shipment at competitive prices, competence of employees and businesses in the area of logistics, possibility of identifying and tracking shipments, and timely delivery of shipments in compliance with the planned lead time. According to the data presented by the Central Statistical Office of Poland (GUS) and the LPI Report, last year's economic sentiment's indicator in the Polish transport retained its progressive profile, which may testify to the optimism characterizing entrepreneurs from this industry.

The analysis of Poland as compared to 139 other countries has shown that in 2023 the situation of the TSL sector improved as compared to the previous editions of the Ranking. In 2023, Poland occupied 26th place (an increase by two places in comparison to 2018) with the synthetic result of 3.6 (2018: 3.54, other data in Table 1.1).

Table 1.1 The Logistics Performance Index (LPI) – Poland

RANKING LPI POLSKA	2023	2018	2016	2014	2012	2010	2007
Ranking position	26	28	33	31	30	30	40
Country score	3.6	3.54	3.43	3.49	3.43	3.44	3.04
Customs position/score	25/3.4	33/3.25	33/3.27	32/3.26	38/3.3	34/3.12	38/2.88
Infrastructure position/score	39/3.5	35/3.21	45/3.17	46/3.08	42/3.1	43/2.96	51/2.69
International shipments position/score	38/3.3	12/3.68	33/3.44	24/3.46	22/3.47	35/3.22	52/2.92
Competence of logistic position/score	33/3.6	29/3.58	31/3.39	33/3.47	32/3.3	36/3.26	38/3.04
Tracking and tracing position/score	23/3.8	31/3.51	37/3.46	27/3.54	37/3.32	33/3.45	40/3.12
Timeliness of shipments position/score	21/3.9	23/3.95	37/3.8	15/4.13	19/4.04	2/4.52	40/3.59

Source: The World Bank, LPI Global Rankings 2018. The numbers presented in the table define the position of Poland in the ranking in terms of a given area.

The top positions in the 2023 Ranking were taken by Singapore, Finland, Denmark, Germany, and the Netherlands.

Escalation of the TSL sector is primarily the effect of a growing demand for transport services in the largest EU markets. On the other hand, year 2019 showed a decreasing tendency in the entire industry. Experts claim that the main problems in the TSL sector are primarily aggravating payment backlogs, high labour costs and lack of qualified personnel, including lack of drivers experienced acutely for a number of years. Approximately 95% Polish transport companies are companies from the SME sector which are, unfortunately, sensitive to the variability and instability of the market. This is coupled with significant competition in the domestic market and growing competition on the side of foreign transport entities. This affects the increasing number of bankruptcies.

No one perhaps imagined that the logistics industry could face so many problems in a relatively short period of time. In 2020, it was affected by the COVID-19 pandemic, which caused huge problems in the form of disrupted supply chains and many operational restrictions due to health regulations. The economy has not yet recovered from the pandemic; the spectre of a major economic crisis looms in the horizon, and on top of all this, all hell of war has broken loose across the eastern border. The war is having a significant impact on the transport industry and supply chains. In addition, Polish TSL companies are currently facing various challenges, such as fluctuating fuel prices, shortage of workers, the Mobility Package and service prices.

An increasingly important challenge for the industry is the automation of transport, including the autonomization of vehicles and the automation of logistics processes and warehouse operations. While recent years have seen a decline in investor interest in the autonomization of passenger vehicles, there is a perceived acceleration of activity in the area of freight transport. Driverless trucks represent an opportunity to make significant savings on fuel and wage costs, and to speed up the logistics process (skipping stops).

AI is growing fast and it is becoming important due to advances that allow complex algorithms or software to be used in transportation. AI technologies are beneficial for all kinds of industries, including transportation. However, there are also challenges associated with use of AI in transportation, including safety concerns, cybersecurity threats and job displacement concerns (Biddala, Duffy and Ibikunle, 2023).

1.4 ICT – The Significance of Development of Poland in This Field

ICT is a very broad term, analysed from the standpoint of many areas of science. These technologies are used in all sectors of the economy, and issues related to ICT are the focus of interests of both economists as well as politicians dealing with the economy and business – both in developed and in developing countries. The term ICT includes all communication media, media allowing the storage of information and equipment allowing information processing. In addition, ICT also encompass a range of digital applications and complex IT systems allowing the actual implementation of higher abstraction level data processing and storage than the hardware level. It is thus correct to state that ICT are the circulatory system of the knowledge-based economy, in which the information society is of dominant significance. This significance is confirmed by relevant numbers (ITU Hub, no date).

1. Internet use increases globally and in every region: 5.4 billion people, equivalent to 67% of the world's population, use the Internet. In the Europe, the Commonwealth of Independent States and the Americas, about 90% of the population uses the Internet. Approximately two-thirds of the population in the Arab States and in Asia and the Pacific uses the Internet, in line with the global average. However, only 37% of the population uses the Internet in Africa as of now.
2. Mobile phone ownership is higher than Internet use: Globally, 78% of people aged 10 and over own a mobile phone in 2023. On average, in every region and every income group, the percentage of individuals owning a mobile phone is greater than the percentage of Internet users.
3. An important barrier in the uptake and effective use of the Internet is a lack of ICT skills.

At the national level, the development of innovative technologies in the field of ICT enables a country to take higher ranking positions (Kabus, 2022). The assessment is carried out using a number of indicators, calculated with the respective index system and applied for analysis of problem areas in politics, as well as for monitoring of progress in the field of innovative technologies introduction (Petkova et al., 2019). Factors indicating the development level in terms of ICT include the ICT Development Index (IDI) and the Networked Readiness Index (NRI).

The ICT Development Index (IDI) has been published annually since 2009 and is used to monitor and compare developments in the field of ICT among countries over time. In the 2023 report, Poland was at 14th place (out of 169 ranked states) with high rising by level as compared to the year 2017 (49th place). It was, one of the highest ranked European states (ITU, 2024).

The World Economic Forum's Networked Readiness Index (NRI) measures the propensity for countries to exploit the opportunities offered by ICT. It is published in collaboration with INSEAD, as part of their annual *Global Information Technology Report* (GITR). The report is regarded as the most authoritative and comprehensive assessment of how ICT impacts the competitiveness and well-being of nations. The list is led by United States, Finland and Singapore. According to this report, Poland was at 34th place out of 134 ranks (Portulans Institute, 2024).

Based on the data in these reports, the ICT situation in Poland has improved in recent years, but there is still much to be done. A necessary condition for the improvement of the level of ICT development in Poland is the support of the public administration, which includes the adjustment of legal regulations and the designation of innovative technologies as a priority in state policy.

1.5 Results

The complexity of managing transport and logistics processes is a massive challenge in finding optimal management solutions that meet the requirements of green development. There are questions about management support for transport and logistics processes (Burinskiene, 2023).

The following issues related to ICT will be analysed in detail: access to the Internet in enterprises, employees with Internet access, website, advertising on the Internet, social media, public open data,

data utilization, electronic sales, cloud computing, AI and digital intensity index.

The percentage of enterprises with Internet access exceeded 98.9%, regardless of the economic activities of the enterprise. This confirms the importance of being able to connect to the global network. In 2023, access to the network in the TSL companies was above the average value. For companies in three sectors, the percentage was 100%. It can be predicted that in the near future all industries will reach 100%. The percentage of employees with Internet access in 2023 was 55.3%. Depending on the type of business carried out by enterprises, there is considerable variation in the percentage of employees with such access. In 2023, for the TSL sector, the indicator was 62.4%. The highest rate occurred in entities engaged in activities related to information and communication which was 97.7% (Table 1.2).

Table 1.2 ICT Usage in Enterprises – Statistic Data

	2015	2019	2020	2021	2022	2023
ENTERPRISES WITH BROADBAND ACCESS TO THE INTERNET						
Total	92.7	98.6	98.6	98.5	98.5	98.7
TSL sector	94.3	98.9	98.9	98.1	98.1	99.0
Highest 100% in 2023 (water supply, sewerage, waste management and remediation activities)	97.6	100	100	99.3	99.2	100
Highest 100% in 2023 (information and communication)	98.2	99.8	100	99.3	100	100
Highest 100% in 2023 (repair of computer and communication equipment)	100	100	100	100	100	100
EMPLOYEES WITH ACCESS TO THE INTERNET						
Total					53.5	55.3
TSL sector	–		–	–	60.6	62.4
Highest % in 2023 (information and communication)					97.7	97.7
ENTERPRISES HAVING A WEBSITE						
Total	65.4	70.2	71.3	71.4	–	67.3
TSL sector	56.0	58.0	56.3	56.1	–	51.7
Highest % in 2023 (electricity, gas, steam and air conditioning supply)	89.0	89.0	90.3	91.2	–	93.5
ENTERPRISES RECEIVING ORDERS VIA COMPUTER NETWORKS						
Total	12.4	16.5	17.9	17.0	18.0	–
TSL sector	7.1	8.9	8.9	8.9	11.6	–
Highest % in 2022 (accommodation and catering)	–	38.1	53.6	51.4	46.6	–

(Continued)

Table 1.2 *(Continued)*

	2015	2019	2020	2021	2022	2023
ENTERPRISES RECEIVING ORDERS VIA A WEBSITES OR MOBILE APPLICATIONS AND EDI-TYPE MESSAGES						
Total						
Website					15.9	
Mobile app					3.5	
TSL sector						
Website					11.3	
Mobile app					1.0	
ENTERPRISES PAYING FOR ADVERTISING ON THE INTERNET						
Total	–	–	–	–	–	20.0
TSL sector	–	–	–	–	–	10.3
Highest % in 2023 (information and communication)	–	–	–	–	–	38.4
ENTERPRISES USING PUBLIC OPEN DATA						
Total	–	–	–	19.0	14.6	–
TSL sector	–	–	–	15.6	12.5	–
Highest % in 2022 (electricity, gas, steam and air conditioning supply)				38.6	33.2	–
ENTERPRISES CONDUCTING DATA ANALYTICS						
Total	–	–	–	–	–	18.0
TSL sector	–	–	–	–	–	14.9
Highest % in 2023 (information and communication)	–	–	–	–	–	39.5
ENTERPRISES OUTSOURCING DATA ANALYTICS						
Total	–	–	–	–	–	3.9
TSL sector	–	–	–	–	–	2.8
Highest % in 2023 (electricity, gas, steam and air conditioning supply)	–	–	–	–	–	11.4
ENTERPRISES USING CLOUD COMPUTING PAYED SERVICES						
Total	–	17.5	–	28.7	–	55.7
TSL sector	–	15.05	–	23.8	–	43.5
Highest % in 2023 (information and communication)	–	53.4	–	67.4	–	80.8
ENTERPRISES USING ARTIFICIAL INTELLIGENCE TECHNOLOGIES						
Total	–	–	–	–	–	3.7
TSL sector	–	–	–	–	–	1.8
Highest % in 2023 (information and communication)	–	–	–	–	–	17.6

Source: Information Society in Poland results of Statistical Surveys in the years 2015–2019 and Information Society in Poland in 2023.

In the age of widespread Internet access, every year more and more companies use their website as a marketing tool. Modern websites are becoming increasingly more technologically advanced and fulfil other roles in addition to presentation functions (Table 1.3).

Table 1.3 Enterprises Classified to Particular Levels of Digital Intensity Index (in % of Total Enterprises in a Group

	VERY LOW	LOW	HIGH	VERY HIGH
Total				
2018	56.3	31.3	11.0	1.5
2023	48.5	30.5	17.1	3.9
TSL sector				
2018	68.1	26.4	4.4	1.1
2023	56.9	30.9	11.0	1.2

Source: Information Society in Poland results of Statistical Surveys in the years 2015–2019 and Information Society in Poland in 2023.

Websites make it possible, for example, to place orders, check their status online, post information about job vacancies, etc. In 2023, less than 70% of enterprises had their own websites. There is a great variation in this indicator; the highest was recorded in the sections manufacturing and supplying in electricity, gas, steam and hot water (93.5%) and the lowest in TSL (51.7%). The most frequently used website function, from the type of business, is the presentation of products, catalogs or price lists of products and services. In 2023, 62.0% of companies in Poland used this website function. The TSL sector is below the average 48.4% in this area (Table 1.4).

The Internet is an excellent advertising tool. It is flexible and gives you the opportunity to better define your target audience. It allows you to prepare personalized content based on information obtained from the web. In 2023, one in five companies paid for advertising on the Internet (Table 1.2). The most frequently paid for advertising on the Internet were enterprises from the Information and Communication section (38.4%),

Table 1.4 Facilities Offered by Websites of Enterprises (in % of Total Enterprises in a Group)

	PRODUCT CATALOGUES OR PRICE LISTS	POSSIBILITY FOR VISITORS TO CUSTOMIZE OR DESIGN THE PRODUCTS	ONLINE ORDERING OR RESERVATION OR BOOKING (SHOPPING CART)	PERSONALIZED CONTENT OF THE WEBSITE FOR REGULAR/ REPEATED VISITORS	ADVERTISEMENT OF OPEN JOB POSITIONS OR ONLINE JOB APPLICATION
Total					
2021	66.8	10.7	13.0	6.8	22.3
2023	62.0	8.0	13.0	13.0	19.5
TSL sector					
2021	51.3	4.1	3.5	2.6	17.1
2023	48.4	4.2	4.6	10.4	6.5

Source: Information Society in Poland results of Statistical Surveys in the years 2015–2019 and Information Society in Poland in 2023.

Table 1.5 Enterprises Using Social Media (in % of Total Enterprises in a Group)

	SOCIAL NETWORKS	BLOGS OR MICROBLOGS	MULTIMEDIA CONTENT – SHARING WEBSITES	AT LEAST ONE OF THE FOLLOWING
Total				
2021	44.0	6.8	16.8	45.6
2023	46.7	6.3	20.2	47.6
TSL sector				
2021	32.5	3.7	8.2	33.9
2023	36.1	3.0	9.2	36.2

Source: Information Society in Poland results of Statistical Surveys in the years 2015–2019 and Information Society in Poland in 2023.

the least were from the Transportation and Warehousing section (10.3%) (Table 1.5).

Open data is data of institutions, offices, companies or research institutes that anyone can use. More and more modern products and services are being developed on the basis of open data in Europe and around the world. Open data saves money and time of entrepreneurs. They can use data resources for pursuing their own goals, developing their business or research. In 2022, 14.6% of enterprises used open public data for business purposes (4.4 percentage points less than in 2021). The situation was similar in the TSL sector where use of open data fell to 12.5% in 2022 compared to 2021 by. Open public data was used mainly by companies from the Electricity, Gas, Steam and Hot Water Generation and Supply section (33.2%).

Enterprises are increasingly turning to various digital tools for analysis. In 2023, the percentage of TSL companies using ERP-type software was 28.6 (36.0% for enterprises in general). For CRM and Business Intelligence software, the figures are 21.8% and 5.5% (general entities 28.5% and 10.5%), respectively (Table 1.6). In 2023, almost one in four companies have electronically shared information with their customers or contractors. Information was shared almost twice as often with customers than to contractors (see Table 1.7).

Table 1.6 Enterprises Using Business Software in 2022 (in % of Total Enterprises in a Group)

	ERP	CRM	BUSINESS INTELLIGENCE
Total	**36.0**	28.5	10.5
TSL sector	28.6	21.8	5.5

Source: Data from: Information Society in Poland results of Statistical Surveys in the years 2015–2019 and Information Society in Poland in 2023.

Table 1.7 Companies Sharing Data in 2023 According to the Entity to Which the Data Is Made Available

	CLIENTS OR SUPPLIERS	CLIENTS	SUPPLIERS
Total	**23.8**	20.2	11.9
TSL Sector	17.8	15.8	9.0

Source: Information Society in Poland results of Statistical Surveys in the years 2015–2019 and Information Society in Poland in 2023.

In 2022, the percentage of enterprises conducting sales through computer networks was 18.0%. Electronic sales via websites, mobile apps or online sales platforms were conducted more often than using EDI-type messages. For the TSL sector, the figures are 11.3% and 1%. This means that the use of mobile applications in this sector is extremely low.

Obligation Chamber refers to the use of scalable ICT services via the Internet. Services may include access to software, use of certain computing power, storage of data etc. In 2023, 55.7% of enterprises used paid cloud services. Taking into account the type of business, the most frequent users of this service were entities in the section of Generation and Supply of Electricity, Gas, Steam and Hot Water (84.3%). The lowest interest in this type of technology was shared by the TSL sector (43.5%).

Chatbots, facial or speech recognition systems, and algorithms that position content on the Internet are some of the manifestations of AI in everyday life. The increasing development of this field of knowledge – related to neural networks, robotics or the concept of artificial life – raises a number of questions not only about its possibilities, but also about the risks it brings. This is a new and difficult technological challenge. Its implementation in enterprises seems inevitable. A quick and accurate decision to apply AI solutions may prove to be a source of competitive advantage.

In 2023, 3.7% of enterprises declared the use of AI technology. Depending on the size of the company and the type of business, there is significant variation in the percentage of enterprises that use AI. The largest share of entities using AI tools was recorded among large entities (24.4%) and in the Information and Communication section (17.6%). In the TSL section, this share is only 1.8% of enterprises (Table 1.8).

Table 1.8 Enterprises Which Incurred Investments on Selected Type of ICT Equipment in 2022

	PURCHASES OF IT AND/OR TELECOMMUNICATION GOODS	PURCHASES OF IT GOODS	PURCHASES OF TELECOMMUNICATION GOODS
Total	**35.8**	**32.9**	**18.4**
TSL sector	32.6	29.1	19.8
Highest 100% (water supply, sewerage, waste management and remediation activities	65.7	62.5	34.1

Source: Information Society in Poland results of Statistical Surveys in the years 2015–2019 and Information Society in Poland in 2023.

The digital intensity index is a general, synthetic index that succinctly reflects the degree of ICT use in enterprises across technologies. In 2023, the majority of Polish enterprises were categorized as having very low or low digital intensity. A high or very high level of intensity occurred for 21.0% of enterprises (Table 1.8). Taking into account the type of business, high or very high intensity was most often characterized by entities engaged in information and communication (56.0%). On the other hand, in the TSL sector, only 1.2% have a very high index, 11% have a high intensity, 30.9% have a low intensity, and unfortunately 56.9% have an indicator at a very low level.

1.6 Conclusions

The aim of this chapter is to analyse the use of ICT in Polish enterprises of the TSL sector. The performed analysis allows for answering the questions posed at the beginning of the study. It must be stated that the TSL sector has a large share in generating economic value for the Polish economy. It is an important part of the Polish economy, however, according to the LPI Ranking in 2023, the share of transport and logistics (commodities and people) in Poland was on the 26th position.

In 2023, the percentage of entities with broadband access to the Internet amounted to almost 99%. The percentage of employees with access to the Internet amounted 55.3%. More than two-third of enterprises had a website (67.3%), which was primarily used to present products, goods and services (62.0%). For most of the areas analysed, the TSL industry is below average.

The study has certain limitations due to the data that were analysed. The analysis is based on results of studies conducted by Statistics Poland and presented in *Information society in Poland in 2023*. This study encompassed enterprises with ten or more employees. This means that the situation of ICT implementation in micro-enterprises was not analysed. However, such an analysis may be conducted at a later time on the basis of statistical data published by the institution. This could also be the vantage point for own research by the authors.

References

Abduljabbar, R. et al. (2019). Applications of artificial intelligence in transport: An overview, Sustainability, 11(1), p. 189. https://doi.org/10.3390/su11010189

Arendt, L. and Grabowski, W. (2019). The role of firm-level factors and regional innovation capabilities for Polish SMEs, Journal of Entrepreneurship, Management and Innovation, 15(3), pp. 11–45. https://doi.org/10.7341/20191531

Bahrini, R. and Qaffas, A. (2019). Impact of information and communication technology on economic growth: Evidence from developing countries, Economies, 7(1), p. 21. https://doi.org/10.3390/economies7010021

Biddala, S. C. R., Ibikunle, O. and Duffy, V. G. (2023). Systematic review on safety of artificial intelligence and transportation, in V. G. Duffy et al. (eds) HCI International 2023 – Late Breaking Papers, HCI 2023. 25th International Conference on Human-Computer Interaction (HCI International), Springer International Publishing AG, Cham (Lecture Notes in Computer Science), pp. 248–263. https://doi.org/10.1007/978-3-031-48047-8_16.

Burinskiene, A. (2023). The role of ICT in transport and logistics processes management, Entrepreneurship and Sustainability Issues, 11(1), pp. 251–267. https://doi.org/10.9770/jesi.2023.11.1(15)

Burnewicz, J. (2017). Racjonalność w transporcie, Zeszyty Naukowe Uniwersytetu Gdańskiego. Ekonomika Transportu i Logistyka, (nr 64 Ekonomika transportu: kierunki współczesnych badań), pp. 33–80.

Button, K. (2010). Transport Economics, 3rd Edition. Edward Elgar Publishing Ltd, Aldershot, Hants, England; Northampton, MA.

Czech, A., Lewczuk, J. and Bołtromiuk, A. (2016). Multidimensional assessment of the European union transport development in the light of implemented normalization methods, Ekonomia i Zarządzanie, 8(4). https://doi.org/10.1515/emj-2016-0035

Eskandari Torbaghan, M. et al. (2022). Understanding the potential of emerging digital technologies for improving road safety, Accident Analysis & Prevention, 166, p. 106543. https://doi.org/10.1016/j.aap.2021.106543

Fechner, I. and Szyszka, G. (2018). Logistyka w Polsce: raport 2017: praca zbiorowa. Instytut Logistyki i Magazynowania. Available at: https://repozytorium.bg.ug.edu.pl/info/book/UOG69b8ef6b383b48c381d7a66be1c3b77b?ps=20&lang=en&title=&pn=1&cid=84065 (Accessed: 15.03.2024).

Furlan, A. D. et al. (2020). Advanced vehicle technologies and road safety: A scoping review of the evidence, Accident Analysis & Prevention, 147, p. 105741. https://doi.org/10.1016/j.aap.2020.105741

ITU Hub. (no date). Measuring digital development – ICT Development Index 2023. Available at: https://www.itu.int/hub/publication/d-ind-ict_mdd-2023-2/ (Accessed: 20.02.2024).

ITU Hub. (2024). Measuring digital development ICT Development Index 2023. Available at: https://www.itu.int/itu-d/reports/statistics/IDI2023 (Accessed: 20.02.2024).

Kabus, J. (2022). Managing the distribution process with an example of the FMCG market. Zeszyty Naukowe. Organizacja i Zarządzanie/ Politechnika Śląska, 160, p. 275.

Krawczyk, P. (2018). Wyzwania rozwojowe polskich przedsiębiorstw sektora TSL, Marketing i Rynek, 9, pp. 447–459.

Krawczyk, P. and Kokot-Stępień, P. (2020). The impact of the exchange rate on the financial result of enterprises in the transport sector, Ekonomia i Prawo, 19(1), p. 47. https://doi.org/10.12775/EiP.2020.004

Mclean, S. et al. (2023). Forecasting emergent risks in advanced AI systems: An Analysis of a future road transport management system, Ergonomics, 66(11), pp. 1750–1767. https://doi.org/10.1080/00140139.2023.2286907.

Mężyk, A., Zamkowska, S. (2019). Problemy transportowe miast: stan i kierunki rozwiązań, PWN, Warsaw. 254 pp.', Prace Komisji Geografii Komunikacji PTG, nr 23(3). Available at: http://yadda.icm.edu.pl/baztech/element/bwmeta1.element.baztechd2d83998-126b-45b1-8acf-175738aff6e6 (Accessed: 15.03.2024).

Olejniczak, K. and Dębicka, A. (2021). Logistyka międzynarodowa a uwarunkowania zarządzania małymi i średnimi przedsiębiorstwami branży TSL w Polsce. Wybrane zagadnienia. https://doi.org/10.21008/J.0239-9415.2021.083.07

Osinska, M. and Zalewski, W. (2020). Determinants of using telematics systems in road transport companies, European Research Studies Journal, XXIII, pp. 474–487. https://doi.org/10.35808/ersj/1604

Petkova, L., Ryabokon, M and Vdovychenko, Y. (2019). Modern systems for assessing the informatization of countries in the context of global sustainable development, Baltic Journal of Economic Studies, 5(2), p. 158. https://doi.org/10.30525/2256-0742/2019-5-2-158-170

Petryczka, I. (2017). Wykorzystanie technologii informacyjno- -komunikacyjnych w przedsiębiorstwach z branży logistycznej, Zeszyty Naukowe UPH seria Administracja i Zarządzanie, 40(113), pp. 83–93.

Piersiala, L. (2018). Influence of special economic zones on the investment activities of enterprises, Organizacja i Zarządzanie : Kwartalnik Naukowy, 2, pp. 41–49.

Portulans Institute. (2024). Network Readiness Index – Benchmarking the Future of the Network Economy Available at: https://networkreadinessindex.org/ (Accessed: 20.02.2024).

Rodrigue, J.-P., Comtois, C. and Slack, B. (2016). The Geography of Transport Systems, London: Routledge, p. 440. https://doi.org/10.4324/9781315618159

Taylor, Z. and Ciechański, A. (2017). Deregulacja i Przekształcenia Przedsiębiorstw Transportu Lądowego w Polsce Na Tle Polityki Spójności UE: Deregulation and Transformation Among Poland's Surface-Transport Companies Against the Background of the EU Cohesion Policy. IGiPZ PAN.

Wandelt, S., Sun, X. and Zhang, A. (2023). AI-driven assistants for education and research? A case study on ChatGPT for air transport management, Journal of Air Transport Management, 113, p. 102483. https://doi.org/10.1016/j.jairtraman.2023.102483

2
Innovative and Smart Cities of the Future

AGNIESZKA ULFIK

2.1 Introduction

In order to be able to talk about smart cities, first it is necessary to define the concept. However, this is not easy as, despite its relatively short history, a number of definitions and explanations of the concept have already appeared in the world literature. In 2015, Albino, Berardi and Dangelico made an attempt to organize the definitions of the time (Albino, Berardi and Dangelico, 2015). They listed 23 most important definitions and their publication is the most cited scientific article on smart cities worldwide (Lazaroiu and Roscia, 2012). It is also important to note that the very concept of smart city covers many scientific disciplines and issues, simultaneously. Krysinski listed as many as 64 areas or scientific disciplines, such as IT, engineering, telecommunications, ecology and environmental protection, economics, management, new technologies, geography, public administration, architecture, social sciences, energy, transport, education, construction, public law, communication, sociology, public health, humanities, arts, geography, automation and control systems, materials engineering, psychology, water resources, geology, health sciences, international relations, neuroscience, anthropology, microbiology and biotechnology, cultural studies, biodiversity and many others (Krysiński, 2020).

The issue of smart cities is closely related to urbanization process. Urbanization is a multifaceted process (including social, legal, ecological, economic and spatial) of urban development, creation of new cities, growth of occupied land, change of lifestyles to urban ones and the concentration of population in urban centres (Korenik, 2019).

DOI: 10.1201/9781003465157-2

2.2 Smart Cities and the European Union

At its core, the smart city concept focuses on the use of advanced technologies, including information technology, as well as the activity and creativity of citizens. In the documents of the European Union (EU), initial directive was toward measures aimed at reducing carbon dioxide emissions into the atmosphere and efficient energy management in cities. It was assumed that this was achievable through the use of advanced technologies with artificial intelligence characteristics through which the quality of public services in cities would be improved. Nowadays, the European Commission indicates that smart cities use digital technologies to increase productivity and improve living conditions, as well as to reduce costs and save resources and increase the level of civic participation (Stawasz and Sikora-Fernandez, 2016).

On the European Commission website, we can read that smart cities are using technological solutions to improve the management and efficiency of the urban environment (https://commission.europa.eu/eu-regional-and-urban-development/topics/cities-and-urban-development/city-initiatives/smart-cities_en [2024.02.02]).

Today, the concept of smart city goes beyond the use of digital technologies to make better use of resources and reduce emissions, as the initial European programmes supporting the development of smart cities aimed to do.

This includes smarter urban transport networks, modernized water supply and waste disposal systems, and more efficient ways of lighting and heating buildings. It also means more interactive and responsive city administration, safer public spaces and meeting the needs of an ageing population.

The "Smart Cities Marketplace" has now been established. It has replaced two previous platforms: "Marketplace of the European Innovation Partnership on Smart Cities and Communities (EIP-SCC Marketplace)" and "Smart Cities Information System (SCIS)". It is a large platform that is changing the marketplace to connect cities, industry, SMEs, investors, banks, researchers and many other smart cities stakeholders.

The Marketplace's main areas of cross-cutting operation include:

1. Sustainable urban mobility,
2. Sustainable districts and built environment,

3. Integrated infrastructures and processes in energy, information and communication technologies and transport,
4. Citizen focus,
5. Policy and regulation,
6. Integrated planning and management,
7. Knowledge sharing,
8. Baselines, performance indicators and metrics,
9. Open data governance,
10. Standards, and
11. Business models, procurement and funding (https://commission.europa.eu/eu-regional-and-urban-development/topics/cities-and-urban-development/city-initiatives/smart-cities_en [2024.02.02]).

Smart Cities Marketplace is one of the key initiatives which aims to promote knowledge exchange, capacity building, and the development and implementation of smart city solutions across the EU. The Marketplace's operations are structured by an integrated Explore-Shape-Deal Matchmaking process, which facilitates collaboration and replication of successful smart city projects.

It is deliberately focused on knowledge exchange, capacity building support and the development, implementation, replication and scaling up of smart city solutions. It is organized in following three phases that complement each other:

a. Explore: See and learn what's next
   This phase provides access to the accumulated knowledge on smart cities, including knowledge on related projects and initiatives. It is an ongoing phase that helps you keep an overview of solutions and best practices that have already been successfully implemented and generate ideas for your own projects.
b. Shape: Shape the project and action plans
   Once the project vision has been developed, this phase helps to transform the idea into a solid project that can attract public and private investors. This phase also enables a structured dialogue between all the key stakeholders involved.
c. Deal: Creating relationships and opportunities
   This third and final phase enables one-to-one exchanges between project promoters and members of the financial community to ultimately close deals and fund projects.

The EU has been actively promoting and supporting the development of smart cities. The EU has launched various initiatives and programmes to facilitate the implementation of smart city solutions and foster innovative urban development.

The European Commission has also been instrumental in advancing smart city projects and providing policy recommendations for local, national and EU decision-makers. These recommendations are based on analysis drawn from the data acquired through smart city projects (EU Smart Cities Information System, 2017). In addition, the European Commission has launched a comprehensive research and development programme for cities which is known as the Smart City and Communities (SCC) programme. This programme aims to support the development of innovative solutions and help cities become smarter and more sustainable (European Commission, 2024).

The EU Parliament has also been actively involved in studying and mapping smart cities in the EU. The parliament commissioned a report to provide background information on smart cities, establish a working definition and map cities fitting this definition across EU member states (European Commission, 2024).

In summary, the EU is playing a significant role in promoting and supporting the development of smart cities across Europe. Through initiatives like the Smart Cities Marketplace, the EU aims to facilitate collaboration, knowledge exchange and the implementation of innovative smart city solutions.

2.3 Functional Areas of Smart Cities – Smart City Wheel

When it comes to building a smart city, it is rarely a case of creating a city from scratch. Most often, the urbanization process has been going on for many decades and smart solutions can now be found in almost every urban agglomeration, even if they were not created intentionally. However, the comprehensive and systemic implementation of this idea, in an already functioning agglomeration, involves complex preparation and rethinking of measures. Currently, the literature most often points to six key elements that must be implemented for a city to be considered smart city (Kinelski, 2022; Krysiński, 2020). From these six elements, Boyd Cohen proposed the creation of a wheel, called the "Smart City Wheel" (Gotlibowska, 2018). These key elements are:

1. Smart economy: It is measured by the city's entrepreneurship and productivity, its adaptation to change, labour market flexibility and international cooperation;
2. Smart mobility: It is measured by local and supra-local accessibility, availability of information and communication infrastructure through the development of sustainable innovative and safe transport;
3. Smart environment: It is measured by the attractiveness of the state of the environment, the level of pollution, environmental protection measures and resource management methods;
4. Smart people: This is characterized by the level of qualifications of people, lifelong learning, social and ethnic diversity, creativity, openness and their participation in public life;
5. Smart living: This parameter is measured by existing cultural facilities, living conditions (e.g. health, safety, housing), educational facilities, tourism attractiveness and social cohesion;
6. Smart governing: It is expressed by the transparency of urban governance, public participation, the level of public services and the implementation of development strategies (Stawasz and Sikora-Fernandez, 2016; Tundys, Bachanek and Puzio, 2022).

Each of the listed factors can be considered separately; however, all of them together create a smart city. These factors were also used to create the multi-criteria model "The Smart City Model", created jointly by researchers from Vienna University of Technology, Delft University of Technology and Ljubljana University (Korenik, 2019). Over the years, other concepts assuming other dimensions of the smart city concept have also emerged (Bitkowska and Łabędzki, 2021); however, they have mostly worked to rename or group individual criteria differently. It is worth taking a closer look at all six areas mentioned above.

2.3.1 Smart Economy and Smart Mobility

In the literature, it is most often assumed that the concept of smart economy is closely related to competition, as these cities compete on many levels. They attract people to their markets (labour, services, money) and thus create dynamic opportunities for development. This leads to the conclusion that a smart city economy should skillfully benefit from location potential, while also using renewable energy sources. This can lead to the diversification of methods that reduce operating costs and the costs of gaining a competitive advantage, especially for residents who are entrepreneurs. As a result, the ability to adapt to the changing economic climate and transform in a way that allows them to survive and thrive in a market economy becomes fundamental.

High productivity is achieved by using and combining factors of production on the basis of knowledge. The service sector becomes extremely important here (Korenik, 2019). In the economic sphere, the intelligence of the city is emphasized through the creative use of information and communication technologies (ICT) in the production of goods and services. The environment of operating businesses is also not without significance. Parks, squares, business districts with the right infrastructure provide opportunities for local businesses to grow while attracting customers.

A characteristic feature of the smart economy is high productivity that is achieved by means of knowledge-based pooling of production factors. However, the importance of heavy industry in urban agglomerations is currently declining, and is certainly declining relative to the situation years ago. The service sector is coming to the fore in the economy, including financial services, IT, education, welfare and trade.

The dominance of the service sector in smart cities is controversial due to the predominance of low-skilled services. This is associated with the low productivity of this sector. Therefore, many researchers emphasize the importance and formalization and codification of services through communication and information technologies in a modern and smart economy. As the global costs of delivery and communication via the Internet are very low, more and more products and services are emerging that can be digitized and automated (Stawasz and Sikora-Fernandez, 2016).

In “The Smart City Model”, the factors that define this area are:

1. Spirit of innovation,
2. Entrepreneurship,
3. Economic image and trademarks,
4. Productivity,

5. Labour market flexibility,
6. International cooperation, and
7. Transformation capacity.

Urban structures are being expanded more and more. The reason for this is the occupation of more and more space for various forms of development, starting with residential, followed by industrial and then service development. For city dwellers, it is always important to be able to get to the centre, and also to ensure free movement within the urban area. For this reason, ensuring mobility on a city scale is of particular importance. Mobility and accessibility are essential elements in meeting the needs of both inhabitants (reaching places of work, education, services, culture, recreation etc.) and entrepreneurs and employers, for whom these two elements provide access to the labour market.

Smart mobility is the adaptation of, above all, advanced solutions in the field of information technology, facilitating the use of all mobility services in the city by its inhabitants, with particular emphasis on sustainable, environmentally friendly solutions, e.g. a smart transport and communication system based on advanced technologies. This will enable existing infrastructure to be used rationally, without creating new facilities. The main objectives to be met by intelligent mobility are reduced congestion, emissions, reduced travel time (distance is not as important as travel time), increased travel safety and minimised travel costs.

Smart mobility solutions are first implemented by city authorities in urban public transport services. They pinpoint the location of rolling stock vehicles, thanks to ICT technologies or virtual transport ticketing solutions. Secondly, city authorities or private operators are introducing cooperative economy solutions such as carsharing, i.e. organized, shared use of not only cars, but also bicycles or scooters. The user usually pays only for the time of use or the distance travelled. This service is attractive because it is freely accessible (no discrimination, objective conditions must be met) and it does not require complicated formalities (no need to conclude contracts each time). In addition, the user does not have to pay for the maintenance costs of the vehicle (which are high in large cities) but can still use them! Thirdly, cities have Mobility-as-a-Service (MaaS) concept, i.e. services to meet the transport needs of users from a single application – a single service combining the offers of different carriers, i.e. multimodal mobility. The user selects a starting and destination point, transport preferences and the app shows them alternative travel options. It allows the user to purchase the tickets needed for the journey, rent a car or order a taxi.

Mobility management is an important issue to ensuring accessibility by shaping a sustainable transport system. In practice, many measures can be taken to improve the transport system. In this context, non-investment measures (soft measures) and investment measures can be taken. The first group includes, for example, transport plans, traffic regulation rules and various types of transport infrastructure user charges. Investment measures (extremely cost-intensive) are primarily aimed at increasing the efficiency of existing infrastructure and its development.

Smart urban mobility manifests itself both in traditional terms – as a system of intelligent, sustainable transport – and in contemporary terms – as digital communication. An intelligent urban transport system is based on the use of advanced technologies for the rational use of transport infrastructure and traffic optimization. It integrates a variety of communication, information and measurement technologies, as well as transport network management techniques in order to achieve specific benefits that increase the efficiency of the system.

The application of smart mobility can lead to reduction in road accidents, increase in street capacity, reduction in road maintenance and rehabilitation costs, reduction in road fleet maintenance costs, reduction in exhaust emissions and an improvement in travel comfort.

In "The Smart City Model", the defining factors are:

1. Local transport accessibility,
2. National and international transport accessibility,
3. Availability of ICT infrastructure, and
4. Sustainable, innovative and safe transport systems.

2.3.2 *Smart Environment, Smart Governance*

The area of smart environment includes all measures related to environmental protection, primarily relating to energy management (including the use of renewable energy) and the measurement, monitoring and control of pollution. In addition, the EU documents the point to the need for specific solutions in the building industry and in urban management in general (e.g. a more efficient waste management system, water resources or street lighting control). The role of a high level of environmental education is also stressed, in order to seek to coordinate and optimize as many environmental activities as possible.

It is reasonable to conclude that the spectrum of arguments in favour of environmental protection in urban centres is immense, and the considerations could be compiled in a separate publication. The concept of a smart environment involves environmental techniques and systems that utilize solutions such as optimizing energy consumption, using renewable instead of nonrenewable energy sources, low carbon policies, environmental education or a system for reducing water consumption. Above all, resource-efficient management requires support to stimulate eco-innovation, i.e. forms of innovation that result in (or aim at) significant and visible progress toward sustainable development. The general principles that are part of the smart environment area are the greening of the economy (taking into account environmental protection requirements in all areas of economic activity), pollution avoidance (prevention), economic efficiency (achieving environmental goals at minimum social cost) and openness of information as a basis for building cooperation and decision-making between participants in the development process. The smart city should strive for energy self-sufficiency. It should be noted that the public, as a group of stakeholders in the city's environmental policy, must be environmentally aware and understand the rationale for environmental policy and actively participate in it. Another important stakeholder in environmental policy is the economic players, often themselves producers of environmental pollution. If the legislation introduced is transparent and restrictive, these actors will take action to protect the environment, often generating competitive advantages in the process.

Promoting the use of RES in the smart city comes under the already existing idea of a green and energy self-sufficient city. Currently, the vast majority of world's energy is generated from fossil fuels, contributing to excessive greenhouse gas emissions into the atmosphere and climate change. The use of RES and the optimization of the urban energy economy are also closely linked to the modernization of the energy infrastructure and the implementation of smart grids and smart metering. Modern smart grids are based on the interconnection of all energy market players, creating an interactive system that enables more efficient energy transmission, better identification of risks, faster restoration of the grid after failures, control of energy demand at different intervals, integration of all sources of energy generation and real-time acquisition of information about the grid and its users.

Energy optimization measures in the smart city also include the implementation of city lighting projects, such as intelligent street lighting based on LED technology and allowing light levels to be adjusted to current weather and traffic conditions.

In "The Smart City Model", the factors defining this area are:

1. Attractiveness of natural conditions,
2. Environmental pollution,
3. Environmental protection, and
4. Sustainable approach to natural resource management.

Smart government is characterized by transparent city management through cooperation (social participation), as well as communication between city authorities and its inhabitants; and ICT is widely used in these processes. Efforts should be made to continuously improve the quality of city management (as part of the principle of self-improvement). To implement the concept, the main requirements, among others are: creating an effective communication system, continuous contact between residents and the city authorities, developing procedures for cooperation between the authorities and social and local groups, and, most importantly, real influence of residents on the way the city functions. Smart city management also fits into the concept of a knowledge-based economy. Thanks to intelligent management, the democratic system and the quality of public service provision are improved and strengthened. It should be borne in mind that intelligent management is accompanied by specific problems, different from the traditional management model. First of all, taking into account activities in the Internet environment, there is a real risk of a hacker attack threatening government structures and creating information chaos in an urban center. Therefore, there is a need to guarantee security through the authenticity of protocols, encryption, authorization and authentication of activities in electronic systems (e.g. in the e-voting system).

Intelligent co-governance results from the integration of the activities of public authorities, social organizations and residents within the city, so that the city can be managed efficiently. This means implementing activities based on public-private partnership and cooperation between various stakeholders. Advanced technologies ensuring interoperability and data exchange are a tool that integrates entities in joint activities. In addition, ICT supports public administration in achieving goals related to its transparency, sharing open data, reducing the costs of its operation and saving time for the implementation of public matters.

The goal of intelligent co-governance in the city is the principle of open government, i.e. full and open access to public information. It allows citizens to be involved in the governance process. The principle of open government together with the concept of the state as a platform provide the basis for the creation of efficient public administration operating within an open platform that creates an environment for the creation of new, innovative services for city users. Also noteworthy is the idea of open data, which allows public data to be opened and made available in digital and standard formats, which enables their reprocessing by residents and encourages entities to create new services, paid or free, on their basis – private complementary public services.

Another goal in this area is the effective provision of public services in the field of the health care system, protection of consumer rights, wide access to legal and consular services, standardization and management of public services, including digital technologies, universal access to the Internet and building efficiently functioning public registers.

In "The Smart City Model", the factors defining this area are:

1. Participating in decision-making processes,
2. Public and social services,
3. Transparency of forms of governance, and
4. Political strategies and prospects.

2.3.3 Smart People, Smart Living, Generation of Smart Cities

Another characteristic of a smart city relates to its users. It is assumed that the higher the qualification level of the city's inhabitants, the greater is the cultural and social diversity; the more open and creative the urban community is, the higher is the standard of living in the city. A key role in this respect is played by education which increases and equalizes employability. No less important is the development of advanced technologies positively influencing the social and economic area of the country, by transforming existing systems of employment, governance, education, production and services.

The ability to use these technologies in everyday life is therefore important. Studies on the development of the Internet show that a significant part of people's economic and social life has moved to virtual reality. Similar changes have taken place in the economic area, as indicated, for example, by the ever-increasing number of commercial transactions concluded online or administrative matters carried out electronically.

In the area of intelligent social and human capital, on the other hand, an important attribute of the society in which this type of capital predominates is the knowledge of individuals – citizens themselves, treated as a potential that, if well managed, results in an increase in the stock of this capital. Knowledge, together with skills and motivation, constitutes the three components of creativity. Creativity is the foundation of economic growth. In the so-called creative economy, a new resource, namely creativity, is becoming a new factor of production and a new social group is also emerging – the creative class, which is the main driving force behind the economy of the post-industrial city.

Important features of intelligent social and human capital are also the absence of divisions in society and tolerance. These features are particularly important in the case of multicultural countries. Another element of this capital is its awareness that the environment and natural resources are its main assets and are essential for the survival and growth of a smart economy. One of the most important binders of social capital is trust, which enables groups organized around certain ideas to develop and achieve their goals efficiently. At the same time, a smart society is eager for continuous lifelong learning through e-learning tools. Therefore, continuous access to education and training becomes a condition for improving the quality of human capital.

Social capital influences innovation processes, as it is one of the components of economic culture, reducing the sense of risk and through this channel it stimulates entrepreneurial and innovative attitudes, determines innovation processes and underpins the modernization of public institutions.

In "The Smart City Model", the factors defining this area are:

1. Qualification level,
2. The ability to undertake lifelong learning,
3. Social and ethnic pluralism,
4. Flexibility,
5. Creativity,
6. Cosmopolitan orientation, and
7. Participation in public life.

The previous area is closely linked to the characteristic of intelligent living. Its determinant is quality of life, understood in a multifaceted way, among other things, as the degree of satisfaction of needs in predetermined areas of life. It is, therefore, the level of satisfaction obtained from the consumption of goods and services purchased from the market, the consumption of public goods, satisfaction with leisure activities and other features of the environment in which we find ourselves.

When relating this concept to urban life, it is important to bear in mind the objective conditions, which include economic and housing conditions, public safety, leisure opportunities, health conditions and the environment. Wide access to public services, social and technical infrastructure, an adequate state of the environment, high level of public safety and having an adequate cultural and leisure offer are therefore determinants of the desired quality of life and characterize a smart city.

The determinant of the sphere of smart living conditions is the quality of life, understood in a multifaceted way. It consists of the level of satisfaction obtained from the consumption of goods and services purchased from the market, the availability of public goods, satisfaction with ways of spending leisure time and with other features of the environment in which we find ourselves. It is worth mentioning that quality of life is identified mainly with social well-being, which is relative in nature and in practice is a difficult phenomenon to measure. Intuitively, however, it is possible to point out that such a city should offer a rich cultural and housing offer, provide wide access to ICT infrastructure for creating lifestyles, consumption and behaviour, and public services should be provided at a high level. Smart living is also associated with strong social cohesion and public participation (public participation in governance). Importantly, smart cities are collaborative endeavours between citizens, authorities, local businesses and other stakeholders, exploiting the rich diversity of roles they play. Such a city cannot be created simply by implementing innovative systems into its space. Smart solutions must correspond to the needs of the city's users.

In "The Smart City Model", the factors that define this area are:

1. Cultural facilities,
2. Health conditions,
3. Personal safety,
4. Quality of housing stock,
5. Educational facilities,
6. Attractiveness for tourists, and
7. Social cohesion.

There are three generations of smart cities (Kinelski, Mucha-Kuś and Makieła, 2022). In line with current trends, all three generations use various types of technologies with artificial intelligence features. Differences occur in the selection of solutions and the group deciding on the choice of solutions used (Korenik, 2017).

In the first one, "Smart City 1.0", advanced technologies play a basic role, but the solutions offered are not adapted to the diverse characteristics of cities. This solution is inspired by available technologies. City authorities are most often not aware of the effects of the implemented technology, and they do not consult these issues with residents or examine their expectations. This is an early phase of the development of smart cities, where the initiators of the use of new technologies are ICT sector companies that offer ready-made solutions to cities regardless of demand. Assessment of the advisability of such investments, their justification and type is not possible due to the lack of qualifications and professional knowledge of people deciding on the purchase, which is often used by companies fulfilling orders.

In "Smart City 2.0", the second-generation smart cities, city authorities play an important role. They are the active party and look for solutions that will work best in the given city. For most cities, the lack of appropriate governance arrangements appears to be the most serious obstacle to their effectiveness and transformation into smart cities. It is a city with a decisive role of public administration.

The latest generation technologies support the implementation of the strategic intentions of the city authorities and contribute to improving the quality of life of residents. In this phase of smart city development, local authorities are actively looking for solutions and striving to improve the quality of life of residents. Unlike Smart City 1.0, here we have a more active search for necessary solutions, and what is important is the end result resulting from the use of modern technologies.

In the third-generation smart city, "Smart City 3.0", the city residents take the initiatives. They are the ones who propose new solutions. Their ideas, expressed in the form of needs and expectations, are translated by modern technology providers into implementable projects. The city authorities play the role of an assistant, observer or support the communication process. Residents often have a part of the city budget to spend in the form of a participatory or citizen budget. The involvement of residents cannot be one-off, they are to create a community of people creating new ideas and solutions. This is the most advanced phase of the development of modern cities, taking advantage of the potential of its inhabitants. This concerns encouraging citizens to use modern technologies and giving them the opportunity to participate in the development of the city through co-decision-making. Cities of this generation, in addition to using the latest technological solutions, focus their activities on social, equality, educational and ecological issues (Bitkowska and Łabędzki, 2021; Krysiński, 2020; Ochojski, 2022).

2.4 Indicators and the ISO 37120 Standard as a Certificate of Being SMART

In recent years, many researchers have been dealing with the issue of smart cities. The relationships between various factors were also examined. Many attempts at measurement have been made and various types of rankings and models have been proposed (Hajduk, 2020; Ryba, 2017).

Scientists all over the world are dealing with these issues, and although this issue is relatively young, the literature on this subject is developing at an avalanche pace. An attempt to assess the amount of research and the connections between them showed that most peer-reviewed publications on smart cities are currently being written in Italy, China, Spain, the United States, the United Kingdom and India (Szczech-Pietkiewicz, Radło and Tomeczek, 2023).

The first recognized publication on the methodology for the Smart Cities Model was by Lazaroiu and Roscia (Lazaroiu and Roscia, 2012). They proposed a system based on a weighted indicator and fuzzy logic.

They proposed a model that allows certain estimates to be made in order for European cities to apply for funds from the EU. This model is consistent with the six areas of smart cities operation also included in the "Smart City Wheel".

In the earlier part of this chapter, the components included in "The Smart City Model" were described in all six categories. Another example of an index is the "Smart Index" proposed by Szczech-Pietkiewicz (Szczech-Pietkiewicz, 2015).

This proposal includes the appropriate components of each category and the weights assigned to them. Each of the six sub-indicators has an equal weight of 1/6.

The components for the sub-indicators are as follows:

Economy

1. GDP PPS per capita (EUR) (25%).
2. Number of registered enterprises/1,000 inhabitants (25%).
3. Activity rate expressed as a percentage (25%).
4. Unemployment rate expressed as a percentage (25%).

People/human capital

1. Median age of the population (33%).
2. Replacement rate expressed as a percentage (33%).
3. Number of students (5–6 grade on the ISEAD scale)/1,000 inhabitants (33%).

Management

1. Public administration is helpful in the opinion of residents (indicator ranging from 0 to 100) (33%).
2. The city's resources are used effectively in the opinion of residents (indicator ranging from 0 to 100) (33%).
3. Res+idents are satisfied with public spaces in the city (index ranging from 0 to 100) (33%).

Mobility

1. Multimodal accessibility (EU-27 = 100) (33%).
2. Number of cars registered/1,000 inhabitants (33%).
3. Residents are satisfied with public transport (indicator ranging from 0 to 100) (33%).

Environment

1. Population density (people/km^2) (33%).
2. Number of days per year when ozone concentration exceeds 120 μg/m^3 (33%).
3. Air pollution is a big problem in the opinion of residents (indicator ranging from 0 to 100) (33%).

Quality of life

1. Number of households living in municipal apartments/ 1,000 inhabitants (33%).
2. Residents feel safe in the city (indicator ranging from 0 to 100) (33%).
3. According to residents, it is easy to find a place to live in the city at an attractive price (index ranging from 0 to 100) (33%).

The Smart Index survey examined 18 out of 45 selected cities. The highest rated were Bordeaux, Groningen, Rotterdam, Lille, Hamburg, Amsterdam, Paris, Praha, Leipzig, Ljubljana and Berlin (Szczech-Pietkiewicz, 2015).

Another noteworthy index is one of the most recognizable indexes: CIMI (Cities in Motion Index). It is led by professors Joan Enrico Ricart and Pascual Berrone from the IESE Center for Globalization and Strategy at the Spanish Business School. Initially, they analysed the level of development in 174 cities, 79 of which were capitals. Later these numbers increased.

To designate CIMI, nine categories are taken into account (IESE, 2022):

1. Human capital: It shows what prospects the city offers to its people and whether it supports their development and stimulates creativity. It is largely influenced by the ratio of people with secondary or higher education, the number of cultural buildings in the city or the number of top business schools.

2. Social cohesion: In this, indicators regarding crimes, the quality of residents' health, their level of satisfaction and the attitude of a given city toward women are analysed.
3. Economics: It is checked whether the city provides appropriate conditions for business development, whether it has appropriate plans for the future and what its approach to entrepreneurs is. Here, attention is paid to the average salary, purchasing power, GDP, GDP per capita and the ease of starting a business.
4. Management: This element directly concerns the efficiency, quality and capabilities of local authorities. The city's finances, ISO 37120 certificate, research facilities, platform with open access to data and corruption index are assessed.
5. Environment: The CIMI Index also included an assessment of whether local authorities care about the environment, whether ecological buildings are being built and what sources of alternative energy are used. Waste management and water consumption as well as air quality and pollution were also analysed here.
6. Mobility and transport: This directly concerns the quality of infrastructure and whether residents are able to move around the city comfortably and quickly. The following factors are taken into account: the number of vehicle rental services, the city's congestion rate and the number of vehicles per household.
7. Spatial planning: Another aspect that is taken into account when assessing the city.
8. International scope: The city's recognition in the international arena and its influence are assessed here, e.g. the number of McDonald's restaurants are also taken into account.
9. Technology: This shows the level of technological knowledge among residents, the level of technology development in the city and access to the latest solutions.

In the last report from 2022, the highest rated cities were:

1. London, United Kingdom (100.00)
2. New York, the United States (98.25)
3. Paris, France (84.99)
4. Tokyo, Japan (80.30)
5. Berlin, Germany (76.42)
6. Washington, the United States (74.27)

7. Singapore, Singapore (73.33)
8. Amsterdam, the Netherlands (73.03)
9. Oslo, Norway (73.01)
10. Copenhagen, Denmark (71.47)
11. Munich, Germany (71.33)
12. Seoul, South Korea (71.22)
13. Chicago, the United States (70.22)
14. Zurich, Switzerland (69.96)
15. Vienna, Austria (69.20)
16. San Francisco, the United States (69.03)
17. Hamburg, Germany (69.00)
18. Dublin, Ireland (68.42)
19. Rotterdam, the Netherlands (68.40)
20. Helsinki, Finland (68.12)

In the acts of international law, it is worth highlighting the set of ISO 37120 standards, which are commonly called smart certificates, and the full name is Sustainable Cities and Communities, Indicators for city services and quality of life. Initially, there was the ISO 37120:2014 standard, which was later replaced by ISO 37120:2018. These standards are the result of cooperation between the standardization organizations of the EU and the International Organization for Standardization in Geneva, as well as national committees.

The set consists of 100 indicators, grouped into 17 areas which are as follows (Malinowska and Kurkowska, 2018; Midor et al., 2021; Midor and Płaza, 2020):

1. Economy,
2. Education,
3. Energy,
4. Environment,
5. Finance,
6. Fire and emergency response,
7. Governance,
8. Health,
9. Recreation,
10. Safety,
11. Shelter,
12. Solid waste,

13. Telecommunication and innovation,
14. Transportation,
15. Urban planning,
16. Wastewater, and
17. Water and sanitation.

The ISO37120 "Sustainable development of communities indicators for city services and quality of life" defines a comprehensive set of indicators that can be used by cities of all sizes to measure and control the level of development in social, economic and environmental terms.

Experts gathered around this organization have concluded that an intelligent, sustainable, prosperous and flexible city of the future can only be built by implementing goals and making decisions based on reliable and standardized data.

2.5 Conclusions

The smart areas shown are compatible with each other. These areas interpenetrate each other, for example, intelligent mobility guarantees public safety, reduces road traffic congestion, and on the economic level, improves the speed of transfer of people and goods, reducing the costs of this transfer. One of the assumptions of the intelligent environment is to positively influence the air quality, limiting pollution, which in turn, is closely related to the health of society, as these pollutants cause many lifestyle diseases, such as cancers, diseases of the circulatory and nervous systems etc. Smart governance aims to enable greater citizen participation in governments, e.g. by providing access to e-technology.

In summary, it is worth paying attention to the extent to which, if at all, smart city activities affect the quality of life in the city. In a broad sense, the quality of life consists of all the needs related to existence in a given urban environment and the state of social satisfaction resulting from the feeling whether these needs are satisfactorily met. Feelings may equally concern the state of the natural environment, the level of wealth, the sense of safety in public places or the possibility of using culture, education, medical care and recreation.

In economics, quality of life is identified primarily with social well-being. However, prosperity is understood as a state of high level of satisfaction of various living and cultural needs of a society or an individual. The most important indicator for assessing prosperity is the level of gross national product per capita. From an economic point of view, the level of prosperity in the city depends on individual and social consumption, including, among others, access to culture and education as well as recreational, housing and employment opportunities.

The quality of life can be considered not only from an economic perspective, but also from a social, subjective one. Without going into further explanation of the concept of quality of life itself, it can be assumed without conducting broader analyses that undertakings, projects and activities in line with the assumptions of the smart city concept fully contribute to improving the living conditions and well-being of city resients. This is also due to the fact that without modern devices, innovative solutions used by intelligent people (entrepreneurs, employees and residents) in business entities and the development of creative industries, progress toward innovation in cities would not be possible.

References

Albino, V., Berardi, U. and Dangelico, R. M. (2015). Smart cities: Definitions, dimensions, performance, and initiatives, Journal of Urban Technology, 22(1), pp. 3–21. doi: 10.1080/10630732.2014.942092

Bitkowska, A. and Łabędzki, K. (2021). Koncepcja inteligentnego miasta — definicje, założenia, obszary, Marketing i Rynek, 2021(2), pp. 3–11. doi: 10.33226/1231-7853.2021.2.1

EU Smart Cities Information System (2017). The making of a smart city: best practices across Europe, European Commission, p. 256. Available at: www.smartcities-infosystem.eu.

European Commission (2024). The State of European Smart Cities, (January).

Gotlibowska, K. (2018). Propozycja modelu miasta inteligentnego (Smart City) opartego na zastosowaniu technologii informacyjno-komunikacyjnych w jego rozwoju, Rozwój Regionalny i Polityka Regionalna, 42, pp. 67–80.

Hajduk, S. (2020). Smart City model and urban spatial management, Gospodarka Narodowa, 302(2), pp. 123–139. doi: 10.33119/gn/120626

IESE (2022). IESE Cities in Motion Index 2022. Available at: http://www.iese.edu/research/pdfs/ST-0396-E.pdf?_ga=1.13056181.367859667.1479064155

Kinelski, G. (2022). Smart City 4.0 as a set of social synergies, Polish Journal of Management Studies, 26(1), pp. 92–106. doi: 10.17512/pjms.2022.26.1.06

Kinelski, G., Mucha-Kuś, K. and Makieła, Z. J. (2022). Koncepcja Smart City i potencjały 4T. Inteligentne zarządzanie miastami Górnośląsko-Zagłębiowskiej Metropolii Dąbrowa Górnicza: Akademia WSB.

Korenik, A. (2017). Smart city jako forma rozwoju miasta zrównoważonego i fundament zdrowych finansów miejskich, Ekonomiczne Problemy Usług, 129(4), pp. 165–175. doi: 10.18276/epu.2017.129-14

Korenik, A. (2019). Smart Cities: inteligentne miasta w Europie i Azji, CeDeWu, Warszawa.

Krysiński, P. (2020). Smart city w przestrzeni informacyjnej. Wydawnictwo Naukowe Uniwersytetu Mikołaja Kopernika, Toruń.

Lazaroiu, G. C. and Roscia, M. (2012). Definition methodology for the smart cities model, energy, Pergamon, 47(1), pp. 326–332. doi: 10.1016/J.ENERGY.2012.09.028

Malinowska, E. and Kurkowska, A. (2018). Norma ISO 37120 narzędziem pomiaru idei zrównoważonego rozwoju miast, Zeszyty Naukowe Organizacja i Zarządzanie/Politechnika Śląska, 2018(118), pp. 363–382.

Midor, K. et al. (2021). PN-ISO 37120 standard – Known or unknown by local administration – Preliminary study, Multidisciplinary Aspects of Production Engineering, 4(1), pp. 489–498. doi: 10.2478/mape-2021-0044

Midor, K. and Płaza, G. (2020). Moving to smart cities through the standard indicators ISO 37120, Multidisciplinary Aspects of Production Engineering, 3(1), pp. 617–630. doi: 10.2478/mape-2020-0052

Ochojski, A. (2022). Miasto inteligentne. Nowe idee, mechanizmy rozwoju, governance. Wydawnictwo Uniwersytetu Ekonomicznego w Katowicach, Katowice.

Ryba, M. (2017). Czym jest koncepcja smart city, a zatem dlaczego powinniśmy je nazywać miastem sprytnym/What is a "smart city" concept and how we should call it in Polish, Prace Naukowe Uniwersytetu Ekonomicznego we Wrocławiu, (467). doi: 10.15611/pn.2017.467.07

Stawasz, D. and Sikora-Fernandez, D. (2016). Koncepcja Smart City na tle procesów i uwarunkowań rozwoju współczesnych miast, Wydawnictwo Uniwersytetu Łódzkiego, Łódź.

Szczech-Pietkiewicz, E. (2015). Smart City – próba definicji i pomiaru, Prace Naukowe Uniwersytetu Ekonomicznego we Wrocławiu, 391, pp. 71–82.

Szczech-Pietkiewicz, E., Radło, M.-J. and Tomeczek, A. F. (2023). Smart and sustainable city management in Asia and Europe, in Dygas, R. and Biswas, P. K. (eds) Smart Cities in Europe and Asia. Urban Planning and Management for a Sustainable Future, Routledge. doi: 10.4324/9781003365174

Tundys, B., Bachanek, K. H. and Puzio, E. (2022). Smart City. Modele, generacje, pomiar i kierunki rozwoju, Edu-Libri, Kraków-Legionowo. Available at: https://commission.europa.eu/eu-regional-and-urban-development/topics/cities-and-urban-development/city-initiatives/smart-cities_en

3

Urban Lab as a Tool to Improve the Quality of Life of City Dwellers According to the Smart City Concept

JANA CHOVANCOVÁ, JUDYTA KABUS AND LUIZA PIERSIALA

3.1 Introduction

Rapid urbanization is a major global challenge. The United Nations Human Settlements Programme (UN-Habitat) reports that cities occupy only 3% of the Earth's land area, but account for 60–80% of energy consumption and 75% of carbon emissions (UN Habitat, 2020). This urban expansion leads to increased pressure on infrastructure, resources and services, exacerbating environmental and socio-economic problems (Chovancová et al., 2023; Seifollahi-Aghmiuni et al., 2022).

In contemporary urban development discourse, the notion of the "smart city" has emerged as a key concept, embodying a holistic approach to urban governance and planning. Embedded in this concept is a commitment to harnessing technological advances, data-driven insights and participatory processes to improve the quality of life of urban citizens while promoting sustainability and resilience (Bibri, 2021). Against the backdrop of unprecedented rates of urbanization worldwide, the imperative for smart city initiatives is underscored by the need to address the complex interplay of socio-economic, environmental and infrastructural challenges that define the urban landscape.

The transition to smart cities is not only innovative but necessary also, aiming to harness technology, data and inclusive governance to make cities more sustainable, resilient and liveable (Lee et al., 2023). This urgency underpins the European Union's (EU) strategic embrace

DOI: 10.1201/9781003465157-3

of smart urbanism, integrating sustainability, inclusiveness and economic vitality into its urban development agenda, and highlighting the critical role of urban labs in pioneering urban transformation.

The importance of cities in Europe, where 78.9% of citizens live in urban areas (i.e. cities and suburbs) and contribute to 85% of the EU's GDP (Eurostat, 2022), underlines the role of cities in driving economic vitality and innovation. This urban concentration is central to addressing Europe's major social and economic challenges, including job creation, growth and investment, as well as leading the way in innovation, energy efficiency, low carbon development and significant efforts to reduce CO_2 emissions.

Within the EU, the journey toward smart cities has been shaped by a confluence of policy imperatives, institutional frameworks and funding mechanisms aimed at stimulating innovation and fostering cross-border collaboration. In particular, initiatives such as the Horizon 2020 programme and the European Innovation Partnership on Smart Cities and Communities (EIP-SCC) have acted as catalysts for mobilizing resources and expertise to advance the smart cities agenda across member states (EIP-SCC, 2016). These efforts reflect a concerted effort to align European urban development strategies with the imperatives of sustainability, inclusiveness and economic competitiveness.

Central to the operationalization of smart city principles are urban labs, which are emerging as dynamic spaces for experimentation, co-creation and knowledge exchange within the urban ecosystem. Designed as collaborative platforms that transcend the traditional silos of government, industry and academia, urban labs embody a new approach to urban innovation based on principles of openness, agility and citizen-centricity (Newton and Frantzeskaki, 2021). Through a plethora of pilot projects, testbeds and living labs, these experimental spaces serve as incubators for novel solutions to urban challenges ranging from mobility and energy to social inclusion and governance.

In this chapter, we embark on a comprehensive exploration of urban labs as key tools for urban transformation in European cities. Drawing on theoretical frameworks and policy analysis, as well as empirical evidence, we seek to map and evaluate urban lab initiatives and approaches and challenges. We aim to distil key insights, emerging trends and best practices that underline the transformative potential of urban labs in enhancing the resilience, liveability and sustainability of European cities within the overarching smart city paradigm.

3.2 Framing the Concept of Smart Cities and Urban Labs

Despite the lack of a universally recognized and accepted definition, the discourse surrounding the concept of smart cities has generated a plethora of interpretations, both academic and practical. Giffinger et al. (2007) emerge as a prominent figure within the discourse, presenting a delineation of smart cities as urban entities characterized by high performance that are dependent on the astute amalgamation of resources and activities orchestrated by self-determined, autonomous and aware citizens who value civic participation and engagement. Dameri and Cocchia (2013) offer an extension of this discourse, emphasizing technological aspects and environmental protection. According to this perspective, a smart city is a geographical area in which advanced technologies such as information and communication technologies (ICT), logistics, energy production, etc., work together to generate benefits for citizens in terms of well-being, inclusion and participation, environmental quality and smart development. It is governed by a well-defined pool of subjects capable of defining the rules and policies for the management and development of the city. Caragliu, Del Bo and Nijkamp (2011) add that the intelligent management of human and social capital, coupled with investment in both conventional and modern infrastructure, not only promotes sustainable economic growth and improves living standards, but also ensures efficient management of environmental resources through participatory governance structures.

3.3 Building Blocks of Smart City Concept

While the definition of a smart city is subject to contextual nuances and interpretative variations, several fundamental attributes collectively characterize its conceptual framework. These are as follows:

1. Technology integration: A cornerstone of smart cities is the seamless integration of advanced technologies across multiple urban sectors, including, but not limited to, transportation,

energy, health care and governance (Kuru and Ansell, 2020). This integration facilitates the collection, processing and use of real-time data, enabling cities to optimize resource allocation, improve service delivery and increase operational efficiency.

2. Data-driven governance: Smart cities rely on the strategic use of data as one of the cornerstones of governance, facilitating evidence-based decision-making and urban planning (Sebastian, Sivagurunathan and Muthu Ganeshan, 2018). By harnessing data from a range of sources, such as sensor networks, Internet of Things (IoT) devices and citizen feedback mechanisms (Sarker, 2022), cities can gain actionable insights into emerging trends, patterns and challenges, enabling informed interventions and tailored solutions.
3. Sustainability and resilience: Cities have become important arenas for addressing sustainability issues, particularly in response to climate change (Toli and Murtagh, 2020). Integral to the smart city ethos is an unwavering commitment to sustainability and resilience. By implementing initiatives that include the deployment of green infrastructure, the use of renewable energy and climate adaptation strategies, smart cities seek to mitigate environmental degradation, reduce carbon emissions and strengthen their ability to withstand and recover from disruptions and stressors (Chovancová et al., 2024).
4. Citizen-centric approach: In the evolving landscape of smart cities, especially in times of urban transformation and digitalization, citizen engagement and attention is sometimes overlooked. It is strongly believed that citizens are essential contributors to the regeneration and development of smart cities (Obedait, Youssef and Ljepava, 2019). The principles of citizen engagement, empowerment, participation and co-creation form the basis of the advocacy approach adopted, which recognizes the significant role of citizen input. This approach recognizes the importance of citizens' voices in exerting demand-side pressure on governments, service providers and organizations, thereby promoting a comprehensive

response to citizens' needs. Furthermore, this engagement is seen as crucial in building a trusted and robust relationship with local governments, contributing to democratic legitimacy and transparency. In the dynamic arena of smart cities, the central role of citizen engagement is highlighted, emphasizing empowerment and co-creation for urban innovation. This citizen-centred perspective, which is critical to addressing urban challenges, ensures that smart cities not only advance technologically, but also meet the diverse needs of their residents. The inclusion of community-based research, as discussed by Christofi et al. (2024), enhances this approach by facilitating the direct involvement of local actors in the design and refinement of urban solutions. By fostering inclusive pathways for participation and co-creation, smart cities are able to empower residents, allowing them to actively participate in the formulation, implementation and evaluation of urban policies and initiatives. This approach cultivates a sense of ownership and belonging within the community fabric, ultimately improving the quality of urban life in a way that is both inclusive and participatory.

5. Economic competitiveness: In the quest to become smart, cities are being recognized for their commitment to enhancing economic competitiveness and fostering innovative ecosystems that are conducive to entrepreneurship, creativity and knowledge sharing. This is achieved through strategic investments in digital infrastructure, human capital development and the facilitation of public-private partnerships, creating an environment that attracts investment, stimulates job creation and catalyzes economic growth trajectories. The important role of smart city initiatives in unleashing innovation and entrepreneurship is highlighted in several studies. For example, Mitra et al. (2023) emphasize the need for a supportive framework for start-ups that includes technology infrastructure and a conducive policy environment to foster entrepreneurship in smart cities, especially in developing countries. Furthermore, Radu and Voda (2022) elaborate on the bidirectional relationship between smart cities and

entrepreneurship, where the former provides fertile ground for sustainable development through access to advanced technologies, skilled human resources and a focus on social and environmental sustainability. Taken together, these perspectives highlight the critical importance of (1) adaptive business models that provide a platform for local authorities, financial institutions, enterprises, SMEs and other relevant actors to work together effectively, and (2) financing, in realizing the potential of smart cities as hubs of economic innovation and growth. Through such synergistic efforts, smart cities are poised not only to strengthen economic competitiveness, but also to ensure the sustainability and resilience of urban development in the face of evolving global challenges.

6. Improving quality of life: Foremost among the objectives of smart cities is to improve the quality of life of residents by optimizing urban systems, services, mobility and promoting social inclusion. This ambition is reflected in studies such as De Guimarães et al. (2020), which highlights the important role of smart governance – through transparency, collaboration and accountability – in improving the quality of life in the Northeast of Brazil. Similarly, Shapiro (2006) highlights the impact of human capital, showing that a higher concentration of college-educated residents in urban areas not only promotes economic growth, but also improves quality of life. Macke et al. (2018) explore this further in Curitiba, Brazil, identifying key areas of quality of life such as socio-structural relationships and environmental well-being, but noting a gap between the city's smart initiatives and residents' satisfaction. Collectively, these studies highlight the need for smart cities to balance technological advances with the development of human and social capital, advocating for a holistic approach to urban planning that prioritizes the well-being and holistic fulfilment of its citizens.

Key components of a smart city are illustrated in Figure 3.1.

The EU has actively promoted the development of smart cities through a variety of initiatives aimed at improving urban sustainability, efficiency and quality of life. These initiatives reflect a comprehensive

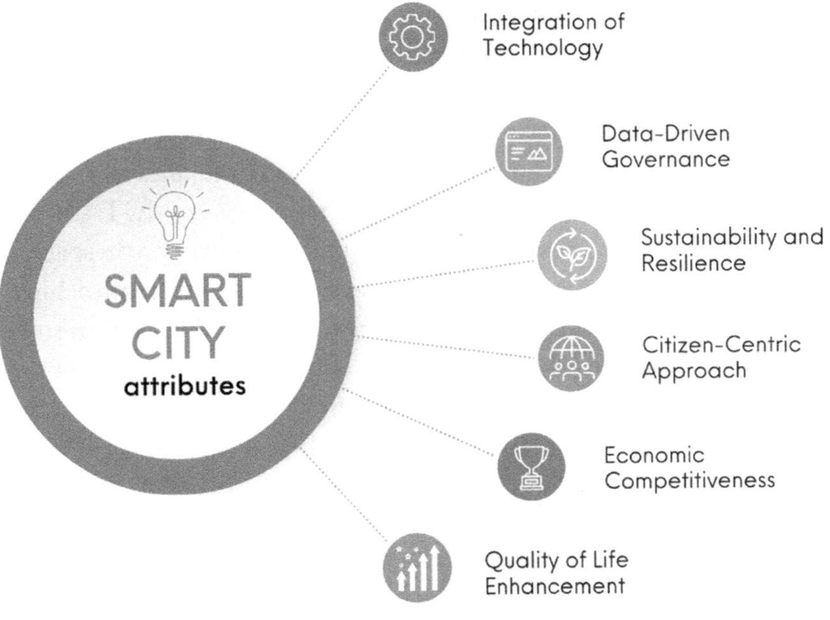

Figure 3.1 Smart city key attributes (own elaboration).

approach to addressing environmental, economic and social challenges through the integration of digital, energy and transport infrastructures. Here are some key EU initiatives on smart cities:

1. Horizon Europe: Following the success of Horizon 2020, Horizon Europe is the EU's main funding programme for research and innovation with a strong focus on smart cities. It aims to promote sustainable growth by supporting projects that integrate digital technologies into urban environments to make cities more liveable, resilient and inclusive.
2. European Innovation Partnership on Smart Cities and Communities (EIP-SCC): The EIP-SCC is a market-changing initiative of the European Commission to support the development of smart cities. It brings together cities, industry, SMEs, investors, researchers and other smart city stakeholders to improve urban life through more sustainable integrated solutions. This includes applied innovation in areas such as sustainable mobility, energy efficiency and the use of ICTs in urban management and planning.

3. Smart Cities and Communities Lighthouse Projects: The Smart Cities and Communities Lighthouse Projects, funded by the EU under the Horizon 2020 programme, embody a strategic approach to promoting sustainable and green urban environments through cutting-edge, scalable and replicable innovations in the fields of energy, transport and ICT. These initiatives are exemplary in their aim to transform urban spaces into models of sustainability and environmental stewardship. At the heart of these projects is the collaboration of a diverse consortium of stakeholders from different sectors and countries. This collaborative effort is aimed at piloting comprehensive smart city solutions that are not only innovative, but also designed to be scalable and replicable in different contexts, thereby maximizing their impact on urban sustainability.

   In this dynamic landscape of urban innovation, the Scalable Cities initiative plays a key role by identifying, promoting and supporting the scaling of integrated solutions and viable business models. This initiative is notable for its extensive network, representing 124 unique cities involved in 20 Smart Cities and Communities (SCC) projects under the Horizon 2020 and Horizon Europe programmes.

   Together these cities have implemented more than 550 demonstrations of technological and social innovations in key urban areas, including mobility and logistics, buildings, urban data and ICT infrastructure, citizen engagement and urban governance. Achievements of Scalable Cities include energy savings of more than 53%, CO_2 emission reductions of up to 88%, the installation of more than 17,500 smart meters, the renovation of more than 1 million m^2 of floor space, the deployment of more than 5,270 electric vehicles (EVs), the installation of nearly 500 electric charging stations and the engagement of more than 260,000 citizens in the transition to a more sustainable urban life (EC, 2024).
4. The Covenant of Mayors for Climate and Energy: This initiative brings together thousands of local and regional authorities who have made a voluntary commitment to implement the EU's climate and energy objectives in their territory. It supports efforts to reduce CO_2 emissions,

increase energy efficiency and use renewable energy sources. Participating cities benefit from the exchange of knowledge, experience and best practices.

5. The Urban Agenda for the EU: The Urban Agenda aims to strengthen the involvement of cities in the design and implementation of policies, including those related to smart cities. It focuses on better regulation, better financing and better knowledge sharing. The agenda includes partnerships on specific issues such as digital transition, urban mobility and energy transition, which are crucial for the development of smart cities.
6. Digital Europe Programme: This programme aims to support the digital transformation of European society and economy. It includes significant funding for digital skills, digital infrastructure and the use of digital technologies in key sectors such as health, education and public administration. Smart cities are a key part of this vision, with investment in high-capacity networks and data spaces that can support urban innovation.

3.4 Urban Labs and Their Role within the Smart City Concept

Urban labs, as part of a broader family of experimental approaches, such as living labs and city labs, have emerged as important ecosystems for innovation in the European urban development landscape. These platforms, highlighted by researchers such as Almirall and Wareham (2011), Kareborn and Stahlbrost (2009), Scholl and Kemp (2016), are dedicated to transdisciplinary research, co-creation and the testing of sustainable solutions. By integrating evidence from this diverse range of experimental labs, urban labs are enhancing their capacity to address urban challenges through a mix of technology, urban planning and inclusive community engagement, seeking solutions that are sustainable, resilient and deeply rooted in the needs of citizens.

Urban labs function as collaborative platforms and are characterized by an inclusive approach to problem solving. They facilitate a multi-stakeholder engagement model, bringing together government agencies, private sector participants, academic institutions and civil society. This mix of perspectives ensures a holistic approach to urban issues, fostering solutions that are not only technologically advanced, but also socially equitable and environmentally sustainable.

Urban labs embody the principles of the smart city concept by focusing on sustainability, resilience and citizen-centred services. They serve as a nexus for innovation, using data-driven strategies and smart technologies to address complex urban issues ranging from traffic congestion and energy efficiency to public health and housing.

Urban labs are central to redefining urban governance and planning by embedding innovation, agility and citizen-centred approaches into their core operations. The key characteristics of urban labs can be outlined as follows:

1. Openness to experimentation: Urban labs are characterized by their experimental nature, serving as test beds for piloting innovative solutions to urban challenges. This openness to experimentation allows for a trial and error approach that is essential to discovering the most effective and sustainable urban interventions. By embracing the possibility of failure as a step toward learning, urban laboratories ensure that new ideas are continuously explored, evaluated and optimized.
2. Agile decision-making: Agile decision-making is another hallmark of urban labs, enabling rapid response to urban challenges and opportunities. This agility is facilitated by the flexible structure of urban labs, which allows them to adapt quickly to changing circumstances and new insights. Such a dynamic approach ensures that urban solutions remain relevant and effective over time, responding promptly to the needs of the city and its inhabitants.
3. Emphasis on citizen-centred design: Urban labs place a strong emphasis on citizen-centred design, ensuring that urban solutions are tailored to the real needs and aspirations of city residents. This focus on user experience encourages the development of services and infrastructure that are accessible, usable and beneficial to the community. By prioritizing citizen perspectives and participation, urban labs promote more inclusive and equitable urban environments.
4. Promoting inclusivity: Inclusivity is at the core of the operational ethos of urban labs, and it aims to involve a broad spectrum of society in the urban development process. This includes marginalized and underrepresented groups, ensuring

that everyone has a voice in shaping the urban landscape. By engaging diverse stakeholders, urban labs work toward creating cities that are not only innovative and sustainable, but also equitable and just.

5. Transparency and accountability: Urban labs promote transparency and accountability in urban governance. By documenting and sharing their processes, methods and results, urban labs are setting new standards of openness in the public sector. This transparency helps to build trust among citizens and stakeholders, while accountability ensures that urban labs remain focused on delivering tangible benefits to the community. Together, these principles contribute to a more democratic and participatory approach to urban development.

3.5 Challenges Faced by Urban Labs and Their Mitigation

Urban labs, while being innovative platforms for sustainable urban development, encounter specific challenges ranging from stakeholder engagement to integrating outcomes with municipal frameworks. This section explores these obstacles, offering insights into how they can be effectively addressed and mitigated to enhance the impact and efficiency of urban labs in driving transformative urban solutions. Drawing on the findings of Scholl et al. (2018), we can outline the main challenges faced by urban labs.

Achieving balanced participation of different stakeholders in the co-design of urban lab experiments is complex due to different interests and expectations. Stakeholders, including citizens, local organizations and government officials, contribute in different ways, which affects the inclusivity of the design phase. While some experiments are initiated by citizens, others are led by officials or researchers, revealing asymmetric stakes and a challenge to fully embrace the co-design. This diversity requires transparent communication and early engagement of all parties to foster a truly collaborative environment for urban innovation.

Aligning around strategic learning goals means defining clear objectives that all stakeholders can understand and support. It requires a shared vision of what the urban laboratory wants to learn and achieve, bridging the gap between different interests. This alignment is critical for focusing efforts on meaningful, transformative outcomes, rather than just operational successes. Establishing these goals early on promotes a unified approach to tackling urban challenges and ensures that the learning process contributes to broader sustainability and governance innovations.

Integrating lab outcomes into local government structures is about ensuring that innovations and learning from urban labs are effectively incorporated into official planning and policy-making processes. This requires establishing robust channels of communication between labs and government agencies, and ensuring that lab activities are aligned with broader city goals and strategies. It is about creating a seamless flow of ideas, results and strategies from experimental spaces into the fabric of urban governance, thereby enhancing policy development and urban planning with fresh, tested insights.

3.6 Examples and Case Studies

Urban labs across Europe are engaged in a wide range of activities that are focused on urban innovation and development. These labs serve as platforms for experimenting with new technologies, policies and social innovations to improve urban life. They typically involve stakeholders from different sectors, including government, academia, industry and the community, in co-creating solutions to urban challenges. Activities can range from pilot projects, testing smart city technologies, workshops and forums for public engagement and idea exchange, to research and development initiatives exploring sustainable urban planning methods. Each lab tailors its activities to local needs and opportunities, fostering a collaborative environment where innovative solutions can be developed and tested in real-world settings.

3.6.1 Innovative Urban Transformation: The Case Study of Košice 2.0

In this section, we present the case of the Košice 2.0 project, which is an example of innovative urban development funded by the European Commission's Urban Innovation Action (UIA) initiative to support sustainable urban growth through experimental projects. It focuses on improving cross-sectoral cooperation, transitioning from an industrial to a digital and creative economy, and strengthening civic engagement. This project aims to improve the quality of urban life, increase civic participation and cultivate cultural and creative sectors for job creation, creating a dynamic ecosystem for innovation and collaboration between local government, citizens and businesses.

The project introduces a creative ecosystem through various initiatives such as:

1. Citizen Experience and Well-being Institute (CXI): CXI plays a key role in the Košice 2.0 project by focusing on improving city services. Its task is to identify, analyse and redesign how these services are delivered, assess their quality and use data to inform policy development and strategy formulation. Integral to its mission is collaboration in the creation of open data platforms. The institute also leads initiatives such as the mobile urban lab and contributes to public art and urban innovation efforts, in line with Košice's UNESCO Creative Cities Network goals.
2. Mobile Urban Lab (MUL): It is a research unit – a point of contact between the urban institute and the city's inhabitants. The MUL takes the form of a car equipped with modern digital technologies, imaging devices and sensors, and it serves to: (1) map public space, the so-called non-places – unused spaces in the city, (2) display data using VR, AR and other modern technologies, (3) identify needs and communicate with city residents, allowing them to participate in defining urban challenges through events, discussions, questionnaires, and (4) implement artistic interventions in public space.
3. Innovation programmes: The Košice 2.0 project creates opportunities for start-ups, entrepreneurs, NGOs and professionals from the IT and creative industries, as well as active citizens of the city, to participate in solving the city's challenges.
4. Bravo Hub aims to create a dynamic space for creatives and start-ups, fostering a community where collaboration and innovation flourish. The aim of Bravo Hub is to create a platform for a community of diverse people to share values, experiences, talents and ideas in a common space. Located in the Cultural Park, its accessible location and comprehensive facilities, including co-working spaces, are designed to inspire and

support the development of original ideas. This initiative underlines the project's commitment to enhancing the city's creative and entrepreneurial ecosystem. Other innovative spaces promoting creativity and collaboration include: (1) the FabLab, a platform offering tools such as 3D printers and CNC machines to bring new ideas to life; (2) the Community & Meetup Space, designed for networking and collaboration between different members of the community; (3) the Audiovisual Centre, which provides high-quality equipment for professional audiovisual productions and supports creative projects.

5. Educational programs within the Kosice 2.0 project are designed to improve the skills of local government, academia and the community. It also uses artistic interventions and media art installations as interactive tools to engage residents, facilitate data collection for the CXI, and enhance urban spaces. These strategies aim not only to encourage community participation, but also to integrate creative and educational endeavours into the city's development efforts.
6. Participatory activities: The Košice 2.0 project includes innovative participatory activities, such as design sprints, which focus on using design methods to address urban challenges in an intensive, abbreviated way. A notable challenge addressed through these sprints is mitigating the effects of urban heat islands in Košice, which significantly affect the quality of life of residents during heat waves. Participants from diverse fields such as ecology, architecture, technology and the arts come together to contribute ideas, learn problem-solving methods and understand the impact of climate change on urban life. This approach fosters collaboration, innovation and community engagement in tackling environmental challenges.
7. Urban Talks: It is a series designed to foster collaboration between the community and the business sector to address local challenges and strengthen the economy. Through these events, stakeholders engage in dialogue to bridge gaps, share perspectives and co-create solutions, with the first talk focusing on regional issues and opportunities for cross-sector collaboration. This initiative underlines the project's commitment to creating an active, innovative ecosystem in Košice.

The Košice 2.0 project case is a transformative initiative that uses European Commission's support to promote sustainable urban development through innovative collaboration and creative solutions. Its comprehensive approach, which focuses on enhancing citizen engagement, growing digital and creative industries and improving city services, positions Košice as a model for smart city development. This case study reflects the project's commitment to building a resilient, inclusive and innovative urban environment, and marks a significant step toward redefining the city's ecosystem.

3.6.2 Urban Lab Częstochowa

The Smart Green City Lab (SGCL) is both a concept and a tool for cooperation between scientific entities, city authorities, residents, businesses and the NGOs, aimed at improving the quality of life of residents through innovative solutions to identified problems and generating additional value using city resources (i.e. initiating, testing, implementing and evaluating projects) (Figure 3.2).

SGCL in Częstochowa is a unit operating within the Częstochowa University of Technology. The aim of the SGCL, established at the

Figure 3.2 Cooperation of actors in the Smart Green City Lab (own elaboration).

Faculty of Management of Częstochowa University of Technology in cooperation with the partner universities, the Housing Management Company Social Housing Association Sp. z o.o. in Częstochowa, Social Economy Entities and the Institute for Urban and Regional Development (IRMiR), is to improve the quality of life of residents. The SGCL is a form of experimental management in which city stakeholders design, develop and test new technologies, products and services to develop innovative solutions to problems arising in various aspects of city functioning. The scope of the Lab in Częstochowa will consist of a wide range of issues, among which the following are detailed: open city data, urban innovation incubator, technological innovation, social innovation, social responsibility incubator and city space management (Kabus and Dziadkiewicz, 2022).

The SGCL is both a concept and a tool for collaboration between scientific entities, city authorities, residents, businesses and the NGOs. The main objective of its activities is to improve the quality of life of citizens through innovative solutions to identified problems and to generate additional value using urban resources (initiating, testing, implementing and evaluating projects).

The SGCL also includes training, workshops, conferences, community activities and educational initiatives.

Five key stages of project implementation are characteristic for the different scopes of SGCL's activities (Figure 3.3).

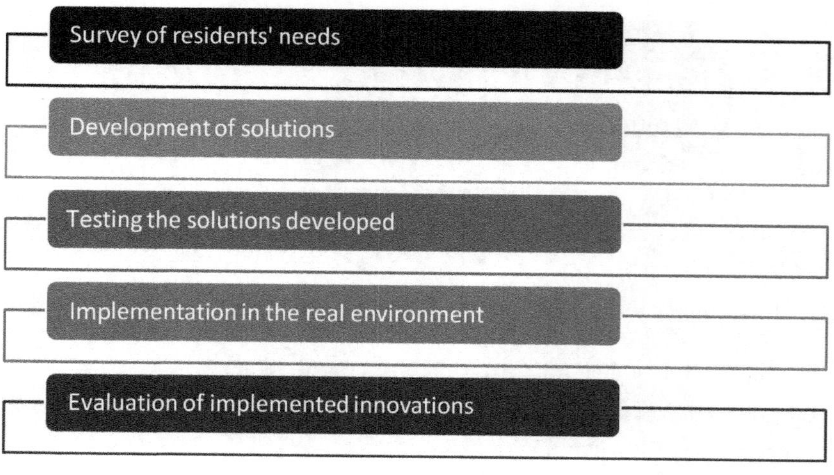

Figure 3.3 Action steps in the Smart Green City Lab (own elaboration).

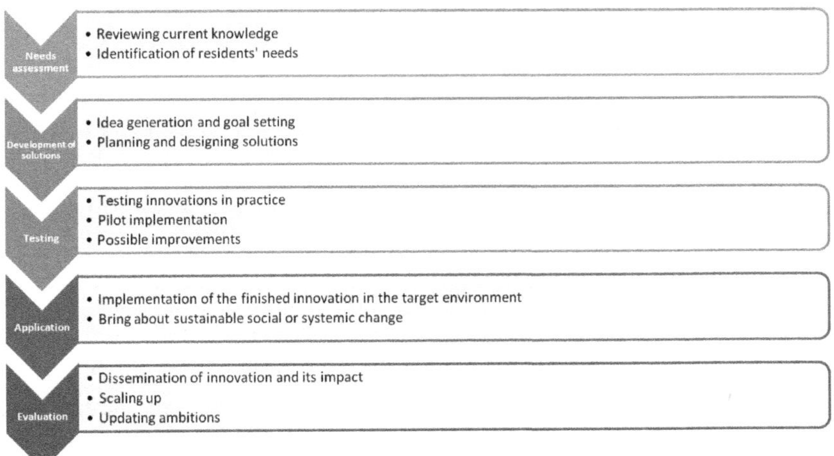

Figure 3.4 Smart Green City Lab project implementation pathway (own elaboration).

Testing/testing can be piloted, on a limited scale (e.g. to one city district, at one intersection or in a selected park), in order to find out about the effectiveness of a solution or its shortcomings.

In a lab, all phases can be implemented or only selected ones for the different types of activities undertaken. The activities undertaken within the lab should each time be concluded with an evaluation of their usefulness for a specific city in the form of an assessment and the possibility of their application on a larger scale (Bień, Jarczewski and Piziak, 2020) (Figure 3.4).

SGCL is also intended to be a place for inspiring meetings, a space for the exchange of modern urban thinking, where numerous thematic events will be organized for and with the participation of local residents and invited guests – experts in their field, from Poland and abroad, who will contribute with their knowledge to breaking the usual stereotypes about how cities function.

Cities developed in line with the smart city concept should be joint ventures between citizens, local authorities, entrepreneurs and other institutions using the variety of roles they can play in urban processes (Bień et al., 2020).

Urban laboratories are not a tool reserved exclusively for city authorities or scientific institutions, but constitute an instrument of close cooperation, synergy of urban stakeholders that are classified into five groups: city authorities (public sector), inhabitants,

NGOs (non-governmental institutions), scientific institutions and business (private sector).

Such collaboration between stakeholders can result in many innovations. Each group of stakeholders has a significant number of different types of resources, both tangible and intangible. On the basis of these resources, new services and products can be produced, ultimately improving the wider quality of life in the city.

It should be emphasized that there are many people, enterprises, organizations and institutions in the market with considerable potential and various types of innovative ideas, the development of which requires, for example, access to city resources or scientific institutions, e.g. in the form of shared city data or expert analyses (JPI Urban Europe, 2013).

By involving different groups of city stakeholders in the activities of the lab, which can offer diverse potentials, the possibility of building a unique urban ecosystem in the form of a lab is created, which will be a model area for prototyping, testing and implementing innovative solutions for the city.

In the presented lab model, unlike other labs, it is not the city council that is in charge, but the scientific institutions are in the role of initiating and managing the institution. Nevertheless, the city's residents are its most important target group. The space of the SGCL gives scientists, students and those involved in research and development projects essentially unlimited opportunities to use the results of their research on a practical basis, and to test and apply them in an urban environment. At the same time, it is not a space limited to the activities of "city" researchers, but is also open to representatives of social sciences, science, technology and many other fields. The Lab creates a space for building interdisciplinary teams (JPI Urban Europe, 2013).

Innovators, start-up owners, entrepreneurs or, ultimately, multinational corporations may also have an interest in getting involved and even financially supporting the Lab, as they themselves are constantly looking for inspiration for their activities and diversifying new ideas for business development or creative employees, who can also be "spotted" at Lab events. Corporate social responsibility (CSR) activities of companies are also not insignificant, the manifestation of which in labs can be the involvement in projects taking into account social interests, environmental protection, as well as the development of relations with different stakeholder groups.

In conclusion, the SGCL should remain flexible so that different stakeholders can freely engage with it at different stages of urban project implementation.

Sources of funding for urban laboratory activities vary widely. Based on a survey of 24 urban labs from different countries around the world, it can be concluded that the type of funding largely depends on the country in which the lab is operating and its policy to support similar instruments. The purpose of the activity (the main theme) and who manages it are also not insignificant (Bień et al., 2020).

With very few exceptions, funding for activities comes from more than one source. Funding can come from public sources (e.g. governmental and non-governmental grants, research grants, funding from international organizations or local governments) as well as from private sources (e.g. funding from commercial companies). Some units also partly sustain themselves through self-financing, i.e. income from their own projects, renting space or organizing paid events (workshops or training for commercial entities). However, grants from the public sector are the predominant mode of funding. It is worth noting that whether a unit was established by the authorities, an academic institution, business or local residents does not fully determine its sources of funding.

It is planned that the SGCL will obtain funding for its activities from various sources, primarily from public sources, i.e. research grants, European funds, government subsidies and from the private sector (Figure 3.5). As the Częstochowa University of Technology, Faculty of Management will be the managing entity of the Lab, it is assumed that the infrastructure of the Częstochowa University of Technology will be used.

3.7 Conclusion

As a key element of the smart city framework, urban labs demonstrate the transformative potential of innovative, experimental solutions to improve the quality of urban life. By fostering collaborative, multi-stakeholder environments, these labs address complex urban challenges with sustainable, resilient and citizen-centred approaches.

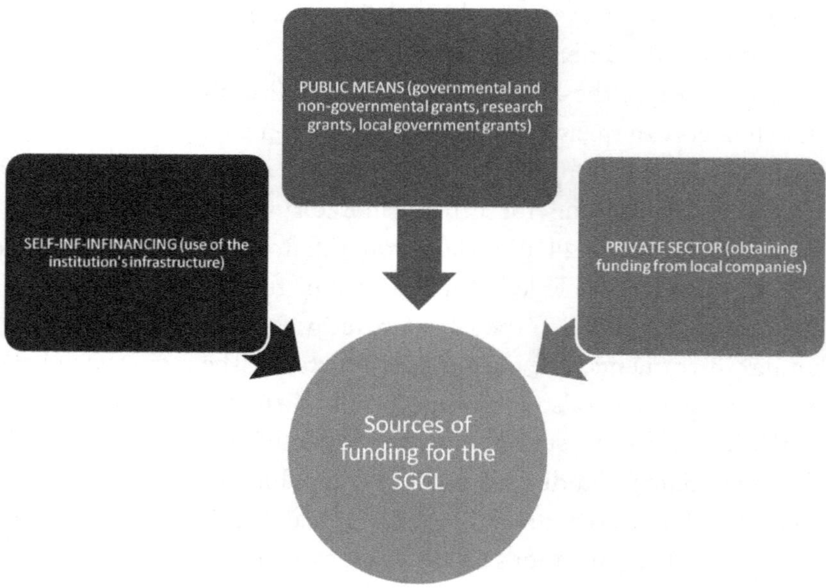

Figure 3.5 Main sources of funding for the SGCL in Częstochowa (own elaboration).

This chapter highlights the importance of urban labs in leveraging technology and community engagement, and advocates for a holistic and inclusive urban development strategy. As cities continue to evolve, the role of urban labs in shaping future urban landscapes will become increasingly important, providing a blueprint for smarter, more liveable cities. In concluding the exploration of urban labs as a tool for improving the quality of urban life, it is important to recognize the importance of managing the diverse interests and expectations of stakeholders. The recommendation to focus on those who are interested in learning about urban challenges underlines the need for targeted engagement in urban labs. This insight is essential for refining the function and impact of urban labs within the smart city ecosystem.

Acknowledgement

This work was supported by the Cultural and Educational Grant Agency of the Ministry of Education, Science, Research and Sport of the Slovak Republic under grant KEGA 010PU-4/2023 and 024PU-4/2023.

References

Almirall, E. and Wareham, J. (2011). Living labs: Arbiters of mid- and ground-level innovation, Technology Analysis & Strategic Management, 23(1), pp. 87–102. https://doi.org/10.1080/09537325.2011.537110

Bibri, S. E. (2021). Data-driven smart sustainable cities of the future: An evidence synthesis approach to a comprehensive state-of-the-art literature review, Sustainable Futures, 3, p. 100047. https://doi.org/10.1016/J.SFTR.2021.100047

Bień, M., Jarczewski, W. and Piziak, W. (2020). URBAN LAB, narzędzie poprawy jakości życia mieszkańców miast zgodne z ideą smart city, Instytut Rozwoju Miast i Regionów, Warszawa-Kraków.

Caragliu, A., Del Bo, C. and Nijkamp, P. (2011). Smart cities in Europe, Journal of Urban Technology, 18(2), pp. 65–82.

Chovancová, J., Petruška, I., Cirella, G. T., Litavcová, E., Chovancová, J., Petruška, I., Cirella, G. T. and Litavcová, E. (2023). Urbanization and CO2 Emissions: Panel Data Analysis of EU Countries in Cirella, G. T., Dahiya, B. (eds) City Responses to Disruptions in 2020. Advances in 21st Century Human Settlements, Springer, Singapore, 123–175. https://doi.org/10.1007/978-981-99-7988-2_8

Chovancová, J., Petruška, I., Rovňák, M. and Barlák, J. (2024). Investigating the drivers of CO2 emissions in the EU: Advanced estimation with common correlated effects and common factors models, Energy Reports, 11, pp. 937–950. https://doi.org/10.1016/J.EGYR.2023.12.057

Christofi, M., Hadjielias, E., Hughes, M. and Plakoyiannaki, E. (2024). Advancing research methodologies in management: Revisiting debates, setting new grounds for pluralism, British Journal of Management, 35(1), pp. 24–35. https://doi.org/10.1111/1467-8551.12791

Dameri, R. P. and Cocchia, A. (2013). Smart city and digital city: Twenty years of terminology evolution. X Conference of the Italian Chapter of AIS, ITAIS, 1–8.

De Guimarães, J. C. F., Severo, E. A., Felix Júnior, L. A., Da Costa, W. P. L. B. and Salmoria, F. T. (2020). Governance and quality of life in smart cities: Towards sustainable development goals, Journal of Cleaner Production, 253, p. 119926. https://doi.org/10.1016/J.JCLEPRO.2019.119926

EC. (2024). Scalable Cities | Smart Cities Marketplace. Available at: https://smart-cities-marketplace.ec.europa.eu/scalable-cities

EIP-SCC. (2016). The Marketplace of the European Innovation Partnership on Smart Cities and Communities (EIP-SCC), p. 27. Available at: https://smart-cities-marketplace.ec.europa.eu/sites/default/files/EIP_Brochure.pdf

Eurostat. (2022). Urban-rural Europe – labour market – Statistics Explained. Available at: https://ec.europa.eu/eurostat/statistics-explained/index.php?title=Urban-rural_Europe_-_labour_market

Giffinger, R., Fertner, C., Kramar, H., Kalasek, R., Pichler- Milanović, N. and Meijers, E. (2007). Smart cities: Ranking of European medium-sized cities.

JPI Urban Europe. (2013). Urban Europe: Creating Attractive, Sustainable and Economically Viable Urban Areas, Joint Call for Proposals 2013, JPI Urban Europe. Available at: https://jpi-urbaneurope.eu/

Kabus, J. and Dziadkiewicz, M. (2022). Residents' attitudes and social innovation management in the example of a municipal property manager, Energies, 15(16), p. 5812. doi: 10.3390/en15165812

Kabus, J. and Dziadkiewicz, M. (2023). Modern management methods in the area of public housing resources in the community, Sustainability, 15(10), p. 7776. doi: 10.3390/su15107776

Kareborn, B. B. and Stahlbrost, A. (2009). Living lab: An open and citizen-centric approach for innovation, International Journal of Innovation and Regional Development, 1(4), p. 356. https://doi.org/10.1504/IJIRD.2009.022727

Kuru, K. and Ansell, D. (2020). TCitySmartF: A comprehensive systematic framework for transforming cities into smart cities, IEEE Access, 8, pp. 18615–18644. https://doi.org/10.1109/ACCESS.2020.2967777

Lee, J., Babcock, J., Pham, T. S., Bui, T. H. and Kang, M. (2023). Smart city as a social transition towards inclusive development through technology: A tale of four smart cities, International Journal of Urban Sciences, 27(S1), pp. 75–100. https://doi.org/10.1080/12265934.2022.2074076

Macke, J., Casagrande, R. M., Sarate, J. A. R. and Silva, K. A. (2018). Smart city and quality of life: Citizens' perception in a Brazilian case study, Journal of Cleaner Production, 182, pp. 717–726. https://doi.org/10.1016/J.JCLEPRO.2018.02.078

Mitra, S., Kumar, H., Gupta, M. P. and Bhattacharya, J. (2023). Entrepreneurship in smart cities: Elements of start-up ecosystem, Journal of Science and Technology Policy Management, 14(3), pp. 592–611. https://doi.org/10.1108/JSTPM-06-2021-0078

Newton, P. and Frantzeskaki, N. (2021). Creating a national urban research and development platform for advancing urban experimentation, Sustainability 2021, 13(2), p. 530. https://doi.org/10.3390/SU13020530

Obedait, A. A., Youssef, M. and Ljepava, N. (2019). Citizen-centric approach in delivery of smart government services, in Al-Masri, A., Curran, K. (eds) Advances in Science, Technology and Innovation, pp. 73–80. Springer, Cham. https://doi.org/10.1007/978-3-030-01659-3_10

Radu, L. D. and Voda, A. I. (2022). The role of smart cities in stimulating and developing entrepreneurship, in Visvizi, A., Troisi, O. (eds) Managing Smart Cities: Sustainability and Resilience through Effective Management, pp. 139–157. Springer, Cham. https://doi.org/10.1007/978-3-030-93585-6_8

Sarker, I. H. (2022). Smart City data science: Towards data-driven smart cities with open research issues, Internet of Things, 19, pp. 1–17. https://doi.org/10.1016/J.IOT.2022.100528

Scholl, C., De Kraker, J., Hoeflehner, T., Eriksen, M. A., Wlasak, P. and Drage, T. (2018). Transitioning urban experiments: Reflections on doing action research with urban labs, GAIA – Ecological Perspectives for Science and Society, 27, pp. 78–84. https://doi.org/10.14512/GAIA.27.S1.15

Scholl, C. and Kemp, R. (2016). City Labs as vehicles for innovation in urban planning processes, Urban Planning, 1(4), pp. 89–102. https://doi.org/10.17645/UP.V1I4.749

Sebastian, A., Sivagurunathan, S. and Muthu Ganeshan, V. (2018). IoT challenges in data and citizen-centric smart city governance, in Mahmood, Z. (eds) Smart Cities: Computer Communications and Networks, pp. 127–151. Springer, Cham. https://doi.org/10.1007/978-3-319-76669-0_6

Seifollahi-Aghmiuni, S., Kalantari, Z., Egidi, G., Gaburova, L. and Salvati, L. (2022). Urbanisation-driven land degradation and socioeconomic challenges in peri-urban areas: Insights from Southern Europe, Ambio, 51(6), pp. 1446–1458. https://doi.org/10.1007/S13280-022-01701-7

Shapiro, J. M. (2006). Smart cities: Quality of life, productivity, and the growth effects of human capital, The Review of Economics and Statistics, 88(2), pp. 324–335. https://doi.org/10.1162/REST.88.2.324

Toli, A. M. and Murtagh, N. (2020). The concept of sustainability in smart city definitions, Frontiers in Built Environment, 6, p. 496662. https://doi.org/10.3389/FBUIL.2020.00077

UN Habitat. (2020). World Cities Report 2020. The Value of Sustainable Urbanization. Key Findings and Messages. Available at: https://unhabitat.org/sites/default/files/2020/11/world_cities_report_2020_abridged_version.pdf

4

Technological Innovations in Transportation

Law and Practice

MARCIN DZIADKIEWICZ

4.1 Introduction

Humankind has always pursued an idea of a perfectly organized society. The beginning of sedentary lifestyle marked the beginning of cities – a new way of organization of peoples. Soon, cities became equivalents of states and their inhabitants started considering them as seats of power – centres of cultural and economic activity. They constituted the main structure that people had a sense of belonging to. Therefore, many have thought on how to improve the lives of city dwellers, as the cities not only created new possibilities, but also new problems that needed to be solved.

The cities grew rapidly and soon Rome became the first city to reach one million inhabitants (Martin, 2012) and therefore it can be called the first metropolis (Blecharczyk, 2022). It was then, just as it is now, an enormous challenge to provide people of a metropolis with proper administration, laws, fresh water and so on. Vast and populous capital, same as vast and populous state, required faster and more effective transportation. Roman engineers put much effort into the design of roman roads, which are considered remarkable to this day (Steiger, 1995). Their innovativeness outlived the state and has been respected for centuries. For it is in human nature that challenges lead to an effort that results in remarkable progress. These days, most interesting and important breakthroughs for city life are the development of information and communication technologies (ICT) and artificial intelligence (AI). The latter is defined as a possibility to learn and an ability to make decisions, which means an ability to utilize gained knowledge by identifying the surroundings and synthesizing "steering" in the most general sense (Choromański et al., 2020). These are the elements,

DOI: 10.1201/9781003465157-4

which nowadays influence the direction of economic and social development, and shape the quite foreseeable future of life on our planet. These changes are mostly visible in cities, for they usually have more resources at their disposal and are economic centres of countries.

The idea of a perfect city is now more up-to-date than ever. According to the United Nations' (UN) data, the urban population will grow from 29.6% of the whole population in the year 1950 to 68.4% in 2050 (United Nations, Department of Economic and Social Affairs, Population Division, 2018). It is a huge growth, given that in 2022 world's urban population totalled 4.54 billion (UNCTAD, 2023), which stood for almost 56.93% of the whole population. The growing numbers of city dwellers and development of new technologies cause new challenges that need to be overcome in order to ensure good quality of life. Therefore, the smart cities idea is thriving and new solutions are being developed nearly every day, for they are greatly desired. Modern technologies, especially as a result of the COVID-19 pandemic, have become increasingly important for the normal functioning of numerous companies, governments or organizations, such as universities (Olejniczak-Szuster and Dziadkiewicz, 2020). At the same time, they have become increasingly important for all of us. This chapter is going to focus on technological innovations in transportation that are implemented all over the world and new laws that have been adopted in order to make the use of these innovations possible and efficient, as well as safe and well-balanced. Some may think that detailed and often late law regulations restrict and limit new ideas, but it is important to set out certain boundaries, because new technologies can be tempting but they can also, in some ways, be dangerous. Therefore, it is reasonable to focus on ensuring the quality of the law, so that it can serve its purpose without causing too much trouble.

4.2 Intelligent Transportation

Intelligent systems are applied in almost every sphere of city life. Transportation is of course one of them. According to the Federal Statistical Office of Germany, there are currently nine cities in the world with intelligent systems with over 20 million inhabitants (Statistisches Bundesamt, 2023). It is much more than the population of most of the countries worldwide. In growing cities with growing population density, efficient and safe means of transport are essential to ensure steady development, popularity among tourists and contentment among residents. In almost all of the cities, cars are intensely used means of transport. Consequently, it comes as no surprise that for instance in the United States, 60% (25,598) of all motor vehicle traffic fatalities in 2021 occurred in urban areas (U.S. Department of Transportation, National Highway Traffic Safety Administration, 2023a). Therefore, there is always a need for safer, more comfortable, efficient and mobile transportation. Another aim is to make our transport ecological, fuel-efficient, material-saving, as well as intelligent and friendly toward persons with reduced mobility.

As we all know, public transport is very effective, providing relatively cheap and safe transportation. But there are also many new areas of transportation that have undergone groundbreaking changes in recent years. Examples like driverless vehicles, autonomous aerial vehicles or delivery drones can be seen as the pictures of the future, although their development had already been intensely sought for years.

4.2.1 Autonomous Vehicles

For years now, the car producing companies have been testing and introducing intelligent systems that improve the comfort of users, functionality of vehicles and safety on the roads. Innovations like the anti-lock braking system (ABS), adaptive cruise control (ACC), electronic brake system (EBS) used for trucks and buses or electronic stability control (ESC) that prevents the possible effects of loss of steering control, have been widely accepted and implemented in numerous types and variants. But the technology does not rest on its laurels and is going forward rapidly. Autonomous vehicles have been in development for some years now and are becoming more and more tangible.

But what do we mean when we say “autonomous vehicles”? There are many different terms that are, sometimes incorrectly, used in reference to this *designatum* – autonomous cars, trucks, buses, motorcycles or even rail and air transport, *platooning*, ADS, ADAS, CAV, CAD, intelligent cars, automated vehicles etc. There are also many different definitions of these. In this section, the theme is autonomous cars. By and large, they can be defined as vehicles that drive autonomously, temporally or constantly, i.e. these eliminate the participation of the driver entirely or partly, allowing to intelligently choose the route (depending on the destination) and perform manoeuvres that are appropriate, considering current situation on the road (Choromański et al., 2020). They should be able to not only turn and brake, but also to change traffic lanes, overtake other cars, drive around obstacles or react to pedestrian crossings. In order to perform these tasks, their systems most importantly need to be able to identify the vehicle's whereabouts and correlate it with an up-to-date road map, distinguish elements of its surroundings using sensor systems and software, make decisions on manoeuvres, implement procedures of safety assessment, as well as solve legal problems (Choromański et al., 2020), in order to comply with the traffic law.

These tasks require lots of elements to be spot on. In order to fulfil these requirements, autonomous vehicles use sensor and information systems. Information systems consist of *software*, *hardware* and algorithms. To function properly, they need to be equipped with access to *big data*, massive computing power and carefully designed algorithms. This IT part requires to be assisted with very precise sensors, using, for example, Differential Global Positioning System (DGPS) (Rahiman and Zainal, 2013), Object and Event Detection and Response or LIDARs – laser scanners (Choromański et al., 2020). When it comes to automated driving, all of these elements basically lead to two main types of systems. Advanced driver assistance systems (ADAS) use data from sensors and systems in order to support human drivers. They feature, for example, ABS or ACC. But there are also advanced driving systems (ADS) that aim to be able to fully operate the vehicle without any participation of the human driver (San Francisco Municipal Transportation Agency, 2019). Though we have to keep in mind that there is no product for sale to the public that is technically fully self-driving on public roads (United Nations Economic Commission for Europe, 2021a) in any circumstances right now. But there is also a more consistent and precise division of autonomous driving systems.

4.2.2 SAE Levels of Driving Automation

It is not (or at least not yet) considered a legal standard, but many official drafts and regulations refer to SAE levels of driving automation that were introduced in 2014 and are widely recognized. They are used in both the press and academic texts, as well as in legal discussions, because they provide a clear division that can be used in the years to come, until full driving automation is reached. This work describes a taxonomy with detailed descriptions of levels of driving automation, ranging from Level 0 (meaning no automation at all) to Level 5 (meaning entirely self-driving systems). There must be a driver in vehicles with Level 0–2 systems, even when the driver support features are engaged, and even if they do not require much action. The driver must constantly supervise these features and he must perform all the manoeuvres that are needed to drive properly. SAE Levels 3–5 mean that there is no need for a driver when the system's features are engaged, whether someone sits in the driver's seat or not, only Level 3 systems require a person to take over when requested. At Levels 4 and 5, systems would not request to do so. When it comes to particular levels, Level 0 includes momentary assistance features like automatic emergency braking, blind spot warning or lane departure warning. Level 1 systems support drivers with lane centring or ACC. Level 2 systems feature both lane cantering and ACC at the same time. When it comes to automated vehicles, the ones that are equipped with Level 3 and 4 systems can drive themselves, but only under limited, specified conditions. At Level 4, pedals or steering wheel may not be installed. Level 5 means full automation – vehicles with systems at this level could operate on their own in any given conditions (SAE International, 2021). Although, such vehicles do not exist yet.

4.2.3 Benefits and Dangers of Autonomous Vehicle Technology

The main goal of the manufacturers of autonomous vehicles is to make these self-driving vehicles perform just like those with human drivers and then to continue to do better. There are numerous benefits that we can derive from this technology. Most importantly, human errors or mistakes – often leading to harm or damages – can be eliminated. The World Health Organization approximated that 1.19 million people die each year following road traffic crashes and pointed out that road traffic injuries constitute the leading cause of death for children and young adults of ages 5–29 years (World Health Organization, 2023). These are startling numbers. Meanwhile, the AI-controlled cars would not speed, would not run a red light, would not force the right of way and finally they would not lose focus or listen to loud music while driving. They can never get intoxicated, drowsy or distracted (Naughton, 2018). They would not lose optimal headway control. Apart from safety, there are also other advantages. The technology most certainly affects businesses too. Many detailed regulations could be omitted, for example there would be no need to apply rest hours, like in the case of bus or truck drivers. The delivery and transport businesses could perform their duties faster and cheaper. Autonomous vehicles could also constitute a groundbreaking change for mobility of elderly people or people with disabilities, as they would not need to steer them. Some say that, thanks to the introduction of driverless vehicles, we could optimize vehicles' work regulations and in effect, reduce environmental harm and improve some road traffic parameters, like for instance travel time or traffic flow capacity (Iwan, 2019).

But there are of course some disadvantages and dangers that come with the use of ADS. The main one is that the system can always be somehow faulty or a malfunction can appear. Also, if the delivery or transport companies begin to use autonomous vehicles in their everyday economic activity, they will be able to continue business with less workers and staff members. Thus, many people in this branch could become unemployed. Also, the advanced technologies and materials used in order to design, program and manufacture self-driving vehicles affect their prices greatly. Therefore, many people cannot yet afford to buy such a vehicle and make use of its benefits. Another one of the concerns resulting from the use of AI in car systems is the impossibility to predict the future development of this branch. Computer science and autonomous cars technology are subjects to intense research and therefore significant adjustments are introduced every other year. There is no specific limit on the research at the moment (Iwan, 2019). As cars assume more control of the driving process than the driver, some new problems can emerge. Firstly, self-driving vehicles are not completely safe yet. There are strict rules on their tests, but accidents sometimes do happen. The first known fatality involving an autonomous vehicle took place in 2018. A pedestrian, who was crossing the road, was hit by an autonomous Uber car. The car system was not designed to anticipate pedestrians stepping onto the road outside crosswalks, but it is a fact that it did not take any action to prevent the accident (Levin and Wong, 2018). And if situations like this are possible, we can never assume autonomous cars being fully safe. However, it is noteworthy that in this case the National Transportation Safety Board concluded in its investigation, that the accident was "avoidable", if the safety driver was alert enough (Riess and Sottile, 2023). The current technologies in vehicles for sale to the public do not enable automated driving systems without any supervision. It will be possible for the driver to take a nap during a car drive only with systems on Level 4 of driving automation. But those are still undergoing tests and research.

Aside these problems, the future aims are of fully autonomous cars that do not need human assistance. But there is still a lot of work to be done. Overall in the United States, according to the U.S. Department of Transportation Information, vehicles equipped with ADS were involved in 492 accidents between July 2021 and October 2023 (U.S. Department of Transportation, National Highway Traffic Safety Administration, 2023b). It is worth pointing out that the number of crashes is growing with time and by far the highest monthly crash rate was reached in August 2023, which is of course related to the increase in the number of cars on road that are equipped with ADS. In the same period, there were 1,177 accidents involving cars equipped with Level 2 ADAS – most automated of the ADAS systems (SAE International, 2018) – in the United States (U.S. Department of Transportation, National Highway Traffic Safety Administration, 2023b).

Moreover, some researchers have been pointing out possible threats concerning the automated vehicles' cybersecurity. In 2015, it turned out that it is possible to hack a connected car driving down the highway in order to control the car's wipers, audio, steering wheel or braking (Miller and Valasek, 2015). In the following years there were many other tests that have proven that security threats concerning connected vehicles are serious and real. For example, in 2018 it was proven that a person can – among other things – hack the steering system of a Tesla autonomous car and control the system with a wireless gamepad or cause the vehicle to divert into the reverse lane by placing some special stickers on the road (Tencent Keen Security Lab, 2019). Usually, these cars are like a closed environment that only acknowledges remote control commands from a single, designated communication channel, but recent autonomous vehicles share their signal and their data with controllers inside the car, as well as external smart and data-processing devices (Park and Choi, 2020). Therefore, there need to be some improvements made in this regard as well.

4.3 Legal Aspects

With the appearance of autonomous cars, numerous new legal problems have shown up. The main ones concern liability assignment, safety regulations including cybersecurity, future requirements for testing of these vehicles, data transfers and software reliability, as well as relations to people. One of the most interesting concerns is the question of liability, i.e. who and in what way is responsible when such a car causes some damages or breaks traffic rules. Under previous regulations it could not be clear. Now it can be presumed that manufacturer liability should increase, as opposed to the driver's liability. On the other hand, should the manufacturer be fully liable? It is not possible to anticipate any possible dangerous situation, even after carefully designing and testing a given technology, therefore strict liability regulations concerning manufacturers could possibly withhold further development of automated vehicles. Of course, to minimize risks, manufacturers should program their autonomous vehicles to cope in extraordinary situations as well, in order to ensure safety of the passengers. That would include not only unpredictable road situations, but also harsh weather conditions or possible vehicle malfunctions. Therefore, regulations concerning testing and certification of autonomous vehicles are first to be adopted. There would also appear new complications when it comes to insurances. These problems are recognized in countries all over the world, which adopt new regulations almost every year.

There are numerous acts that have a worldwide impact on autonomous vehicles and therefore they can only be briefly described. Basic grounds for world regulations concerning vehicles are constituted by the United Nations (UN) Agreements from 1958 and 1998. New UN regulations are being adopted as annexes to these Agreements. Another one of the important international agreements concerning car transportation is the Vienna Convention on Road Traffic from 1968, subscribed to by 89 countries worldwide. This treaty outlines standard traffic rules and aims to harmonize national laws, and to ensure safety of road traffic. However, many developed countries such as Canada, the United States, Japan or China are not parties to this agreement. The Agreement was amended in 2016, allowing autonomous vehicle technologies in road traffic, but under a condition that they comply with the UN regulations concerning vehicles (Choromański et al., 2020).

The need of regulatory activities resulted in the establishment of the Working Party on Automated/Autonomous and Connected Vehicles, called GRVA, by the World Forum for Harmonization of Vehicle Regulations (WP.29), under the auspices of the UN Economic Commission for Europe (UNECE). Its works resulted in adoption of the first UN Regulation on Automated Lane Keeping Systems (ALKS) that approved the use of Level 3 systems that perform automated lane keeping in passenger cars and vans at low speed on motorways (United Nations Economic Commission for Europe, 2021b). In 2021, the approval has been extended for heavy vehicles such as trucks, buses and coaches as well. In the same year, the first series vehicle equipped with an ALKS was launched in the market (United Nations Economic Commission for Europe, 2021a). The World Forum has also adopted a document drafted by the representatives of China, the EU, Japan and the USA. It is called the Framework on Automated/Autonomous and Connected Vehicles (FDAV). It defines a safety vision, key safety elements, programme of activities suitable for countries under the regime of type-approval and countries under the regime of self-certification, as well as provides guidance to the GRVA. The Framework's aim is to harmonize world's automated driving regulations and create better environment for innovation (United Nations Economic Commission for Europe, 2021a). The document is used by the GRVA while working on various projects.

Under the GRVA, an informal working group on Functional Requirements for Automated Vehicles (FRAV) was also established. FRAV focuses on ADS, which means systems capable of driving without human supervision. For example, it defines performance requirements applicable across ADS configurations, which should ensure that ADS have means to perform dynamic driving tasks (DDT), drafts definitions of these requirements and signalizes that manufacturers need to provide descriptions of ADS and its features (FRAV, Informal Working Group on Functional Requirements for Automated Vehicles, 2020). FRAV coordinates its efforts with informal working group on Validation Methods for Automated Driving (VMAD). It was instructed by the GRVA to undertake development of New Assessment/Test Method for Automated Driving (NATM) master document, which outlines a conceptual framework for validating the safety of ADS and guidelines that could provide direction to developers and contracting parties of the 1958 and 1998 UN Agreements on recommended procedures for validating the safety of ADS. The document itself states that it represents current best practices identified by the VMAD for validating the safety of ADS using the NATM and it aims to provide clear direction for validating the safety of ADS in a manner that is repeatable, objective and evidence-based, while remaining technologically neutral and flexible enough to foster ongoing innovation by the automotive industry (GRVA, 2022).

There are also some important UN acts under the 1958 Agreement, such as UN Regulation No. 155 (Cyber Security and Cyber Security Management System), UN Regulation No. 156 (Software Update and Software Update Management System), UN Regulation No. 157 (ALKS) and UN Regulation No. 160 (Event Data Recorder – EDR). In 2023, amendments to UN Regulation No. 157 came into force, allowing the use of assisted lane-keeping at higher speeds, changing the automated driving speed limit from 60 km/h to 130 km/h and stating that an autonomous vehicle is obliged to be equipped with minimum forward detection range of 150 m while driving with maximum speed of 130 km/h (United Nations Economic Commission for Europe, 2022). This document is noteworthy, for it constitutes the first worldwide regulation that concerns Level 3 vehicle automation.

4.4 Selection of Local Approaches

It needs to be underlined that the UN regulations are mostly general, as their goal is not to define precise rules concerning the manufacturing or use of autonomous vehicles. Among the adopted acts there are various guidelines and recommendations that aim to harmonize the laws of particular countries when it comes to autonomous vehicles. National approaches can nevertheless be fairly different. The laws differ when it comes to specific provisions and there are also countries that did not adopt complex statutes that would regulate this matter.

But there are some countries that have passed certain bills concerning autonomous vehicles. Selected approaches are briefly described in subsequent sections.

4.4.1 The United States

The United States, as world's leader in terms of technology and computer science, was one of the first countries to develop and use autonomous vehicles. When it comes to U.S. law it is important to keep in mind that there are federal and state provisions, binding simultaneously. But there are still no complex regulations concerning safety, data privacy or cybersecurity neither at the federal nor state level – in 2017 the House of Representatives and the Senate could not reach agreement when the adoption of the Self Drive Act was discussed (Jones Day, 2021).

However, the U.S. Department of Transportation was not idle regarding this issue. In 2016, it published a document called Federal Automated Vehicles Policy (U.S. Department of Transportation, National Highway Traffic Safety Administration, 2016). It contained guidelines that ought to be respected on the state and federal level. The document has been regularly updated, and in 2017 "Automated Driving Systems: A Vision for Safety 2.0" was published (U.S. Department of Transportation, National Highway Traffic Safety Administration, 2017). The latest document was called "Ensuring American Leadership in Automated Vehicle Technologies: Automated Vehicles 4.0" (U.S. Department of Transportation, National Highway Traffic Safety Administration, 2020). It was based on the previous document, "Preparing for the Future of Transportation: Automated Vehicles 3.0, known as AV 3.0" (U.S. Department of Transportation, 2018) from 2018 and contained a commitment to integrate autonomous vehicles with the whole system of transportation in a safe and gradual way. This integration process was to involve creation of ethical laws for self-driving cars, for it was considered very important to ensure safety of the users of these vehicles, as well as the users of classic, personally steered ones. The AV 4.0 focuses on three main areas. Firstly, the U.S. Government's automated vehicle technology principles that were coined to protect users and communities, promote efficient markets and facilitate coordinated efforts. Secondly, administration efforts supporting automated vehicle technology growth and leadership. And finally, U.S. Government activities and opportunities for collaboration, for example when it comes to research, cybersecurity, infrastructure, taxation or environmental quality. The document aims to coordinate efforts across the federal government and to provide guidance to federal agencies, innovators and the public on the government's posture toward automated vehicles, as well as to ensure a consistent approach to self-driving vehicles technologies across the United States (U.S. Department of Transportation, 2021).

Nevertheless, the most important vehicle regulations on federal level are contained in the Federal Motor Vehicle Safety Standards (FMVSS). These provisions are situated in Title 49 of the Code of Federal Regulations, Part 571, Subpart B. Standards are grouped into categories: crash avoidance, crashworthiness, post-crash survivability and miscellaneous. These regulations originally were meant to apply to regular, manually steered cars, but in recent years a call for a change has appeared. In 2020, the National Highway Traffic Safety Administration (NHTSA) published a notice of proposed rulemaking relating to the development of vehicles equipped with ADS, called Occupant Protection for Automated Driving Systems. It aimed to clarify ambiguities in applying current standards to vehicles equipped with ADS. It was also to help facing challenges of testing and verifying compliance with the FMVSS of these vehicles, because they lack traditional manual controls, necessary for human drivers. In the same year, the NHTSA prepared the Framework for Automated Driving System Safety. It was an advance notice of proposed rulemaking that defined and assessed the safety of ADS-equipped vehicles' performance. Its goal was also to ensure some flexibility while regulating safety issues, in order to enable further innovation. The next document of this kind is the Occupant Protection for Vehicles with Automated Driving Systems, which was published in 2022. It is a final rule that amends the occupant protection of the FMVSS to account for future ADS-equipped vehicles in this regulation. The final rule underlines that vehicles with ADS technology must continue to provide the same, high levels of safety and security of the passengers that current vehicles provide. It aims to clarify existing terminology and thus resolve problems in applying the current standards to ADS-equipped vehicles. In 2023, the NHTSA also amended the Standing General Order 2021-01 that was addressed to manufacturers and operators and concerned their obligation to report crashes of ADS and Level 2 ADAS vehicles to the NHTSA. The process became more formalized in order to fulfil its preventive function.

Many fragmentary regulations concerning autonomous vehicles were adopted on the state level. In most of the states, there are laws or executive orders related to this topic, regulating among other things the question of liability, in a slightly varying manner. For instance, the state laws of Michigan and Tennessee hold manufacturers liable for accidents caused by defects either in the design or construction of the automated vehicles equipped with ADS technology. However, the Michigan law sees manufacturers only as providers of taxi-like services to consumers, whereas Tennessee law addresses privately owned and operated vehicles as well. Moreover, it adds that the liability for accidents will be determined in accordance with product liability law and common law (Quidachay-Swan, 2019).

One of the most active states on this field is California. This state passed legislation allowing automated cars to be tested on roads already in 2012. In the California Vehicle Code, the lawmakers briefly described the legal way of testing and certification of autonomous vehicles, providing definitions of autonomous vehicle ("any vehicle equipped with autonomous technology that has been integrated into that vehicle"), operator or manufacturer. It also obliged the driver to be ready to take over the control of the vehicle. The state of California allowed the first fully self-driving cars (SAE Level 4 or higher) to be tested on public roads under specific conditions in 2020. Soon after, appropriate permits were granted for these cars to transport passengers. Since 2022, self-driving taxis operate in specified locations and hours, mostly in highly developed, urbanized areas and since 2023 at all times of day in San Francisco (Haves, 2023, p. 11). Unfortunately, the technology is not yet perfected. According to the data of California Department of Motor Vehicles, "as of January 26, 2024, the DMV has received 684 Autonomous Vehicle Collision Reports" (California Department of Motor Vehicles, 2024).

4.4.2 The European Union

Different kind of regulation on autonomous vehicles can be seen in the European Union (EU). Although it is once again noteworthy that each and every member state of the EU adopts its own laws and has its own legal system. Some of the member states, such as the Netherlands or Scandinavian countries are one of the leaders when it comes to AI and autonomous vehicle technology. But the EU also has its legal system and therefore its organs pass regulations that are binding for the member states or soft law instruments, such as guidelines or recommendations. The EU has already introduced several legal acts to stay in the loop and improve Europe's competitiveness when it comes to self-driving vehicles.

The year 2016 constitutes the beginning of Europe's journey toward connected and automated driving. In April, the Declaration of Amsterdam was published. In this Declaration, the member states, European Commission and private sector acknowledged the importance of ICT and telecommunication sector development and opportunities related to connected and automated vehicle technologies. These parties agreed to support shared objectives, such as supporting technology development, further innovation or ensuring data protection and privacy. The act presents a joint agenda, which includes topics like coherent European and national rules, use of data, security, international cooperation, vehicle-to-vehicle and vehicle-to-infrastructure communication or common definitions for autonomous vehicles. At the end of the document, member states, the European Commission and industry representatives take on some obligations of action, in order to fulfil the objectives that are stated in the Declaration.

In 2017, the High Level Group on the Competitiveness and Sustainable Growth of the Automotive Industry in the EU (GEAR 2030) published its final report (High Level Group on Competitiveness and Sustainable Growth of the Automotive Industry in the European Union, 2017). The report, it underlines the importance of the automotive industry for Europe's society and economy, analyses the global situation within the market and presents future recommendations concerning the main problems and chances for the automotive sector in the EU. The main conclusions of the report are: competition is growing from non-EU manufacturers in new technologies, development of digital technologies and environmental protection will have a significant impact on the industry, importance of zero-emissions vehicles is increasing, there is a need for a shared strategy on automated and connected vehicles as underlined by the Amsterdam Declaration of

2016 and finally, there will be a significant impact on the workforce of the sector. Among the recommendations concerning specifically automated and connected vehicles there are for instance such as development of large scale open road testing and trials by the Commission and member states, inclusion of data storage requirements in the type-approval legislation to clarify liability, coordination by means of only one EU-wide focal point for open road testing and trials, agreement on regulatory approaches that foster the investments (High Level Group on Competitiveness and Sustainable Growth of the Automotive Industry in the European Union, 2017).

The EU's future plans concerning automated mobility were presented in the Communication from the Commission called "On the Road to Automated Mobility: An EU Strategy for Mobility of the Future", from 2018. The key goal of the EU is supporting the deployment of safe connected and automated mobility, while simultaneously addressing societal and environmental concerns, which are considered to be decisive for public acceptance. The deployment of driverless mobility is expected to lead to achieving no road fatalities in Europe by 2050 – a goal known as Vision Zero. The document refers to SAE Levels of Automation. It is underlined that vehicles equipped with Level 1 and 2 systems are already available and vehicles equipped with Level 3 and 4 systems are being tested. Other elements of the future plans on automated mobility are: allowing innovation, making automated mobility safe, addressing liability issues, fostering vehicle connectivity for automation as well as ensuring cybersecurity, data protection and data access. Though it is emphasized in the Communication that at this early stage it is difficult to foresee all the possible effects of driverless mobility on the transport system, economy, environment or labour market.

In November 2019, the EU adopted Regulation 2019/2144 that amended Regulation 2018/858. Most of its provisions entered into force in 2022, apart from some, which began to bind in 2020. The document, as an EU Regulation is binding in every member state. It is known as Vehicle General Safety Regulation and is fulfilling the plans of the Amsterdam Declaration and Communication from the Commission of adopting an EU-wide legal framework for connected and automated vehicles, which puts safety at the centre of interest. As the Regulation states in Article 1, it "establishes requirements: for the type-approval of vehicles, and of systems, components and separate technical units designed and constructed for vehicles, with regard to their general characteristics and safety, and to the protection and safety of vehicle occupants and vulnerable road users; for the type-approval of vehicles, in respect of tyre pressure monitoring systems, with regard to their safety, fuel efficiency and CO_2 emissions; and for the type-approval of newly manufactured tyres with regard to their safety and environmental performance". The safety measures are expected to save more than 25,000 human lives and avoid at least 140,000 serious injuries by 2038, because EU estimations show that human error is involved in 95% of accidents (European Commission, 2022).

The Regulation introduces several mandatory ADAS to maximize road safety and provides the legal framework for the introduction of fully self-driving vehicles in the EU countries. Some of the provisions apply to all road vehicles (Article 6), but there are also some provisions that were designed specifically for passenger cars and light commercial vehicles (Articles 7 and 8), buses and trucks (Article 9), hydrogen-powered vehicles (Article 10) or automated vehicles and fully automated vehicles (Article 11). The Regulation demands that all motor vehicles should be equipped with the following vehicle systems: intelligent speed assistance, alcohol interlock installation facilitation, driver drowsiness and attention warning, advanced driver distraction warning, emergency stop signal, reversing detection and event data recorder. The legislator provides binding definitions of these terms in Article 3. In regard to automated vehicles, they need to comply with the technical specifications that are set out in the implementing acts and that relate to: "systems to replace the driver's control of the vehicle, including signalling, steering, accelerating and braking; systems to provide the vehicle with real-time information on the state of the vehicle and the surrounding area; driver availability monitoring systems; event data recorders for automated vehicles; harmonized format for the exchange of data, for instance for multi-brand vehicle platooning; systems to provide safety information to other road users".

Since the adoption of Regulation 2019/2144, the Commission has drafted several related implementing regulations, covering various driver assistant measures that were introduced by the Regulation (European Commission, 2022). For instance, in 2022 the Commission Implementing Regulation (EU) 2022/1426 was adopted. The most important parts of the Implementing Regulation are four Annexes. The first one concerns information documents that need to be delivered by a manufacturer of ADS-equipped cars while requesting EU type-approval of fully automated vehicles with regard to their ADS. The second specifies performance requirements, technical specifications, software updates, ADS data requirements etc. The third annex addresses the process of compliance assessment conducted by the authorities, in relation to technical specifications – which factors should be considered. And finally, the fourth one specifies the process of preparation of an EU type-approval certificate of a type of fully automated vehicle, in accordance with the requirements laid down in the Regulation. There is also the Commission Delegated Regulation (EU) 2022/2236 from 2022 which amends Annexes to Regulation (EU) 2018/858 in the sphere of, among others, automated vehicles – mostly their technical specifications, mentioned in the Regulation 2019/2144.

4.4.3 The United Kingdom

In the United Kingdom the need to pass new legislation concerning autonomous vehicles has been seen a few years ago. The most important act in force is the Automated and Electric Vehicles Act 2018. Part 1 of this Act consists of provisions concerning "Automated vehicles" and the question of liability. Section 1 obliges the UK's Secretary of State to prepare a list of motor vehicles that are in at least some circumstances or situations designed or adapted to be capable of safely driving themselves. These vehicles may legally be self-driving on roads or other public places in Great Britain. At present, there are no self-driving vehicles listed for use in Great Britain (Self-driving vehicles listed for use in Great Britain, 2022), as the provision does not apply to vehicles that are only equipped with assisted driving features, with which the driver is still responsible for driving lawfully. Therefore, whole Part 1 of the Act will apply only when a list of automated vehicles is published. But the provisions of this part reflect intentions and objectives of the legislature. Section 2 states that if an insured automated vehicle causes an accident while driving itself and any person suffers damage as a result, the insurer is liable for the damage. But when the same accident occurs with the participation of a vehicle that is lawfully not insured, the owner of the vehicle is liable for the damage. Under Section 3, the amount of the liability of an insurer or vehicle owner can be reduced in an appropriate way in the case of contributory negligence – if the accident was, to any extent, caused by the injured party. But under Section 4 it is possible to exclude or limit the insurer's liability in an insurance policy, if the damage suffered by an insured person arising from an accident is occurring as a direct result of "software alterations made by the insured person or with his knowledge that are prohibited under the policy, or a failure to install safety-critical software updates that the insured person knows, or ought reasonably to know, are safety-critical". Section 5 subsection 1 provision states that when liability of the insurer or vehicle owner to the injured party is settled in respect of the accident, any other person liable to the injured party in respect of the accident is under the same liability to the insurer or vehicle owner – that means a right of insurer or vehicle owner to claim against persons responsible for accident.

Some significant regulations are contained in the Highway Code. It contains certain rules that are applied while driving any kind of vehicle, automated or not. For example, the driver cannot be intoxicated, drowsy or under the influence of certain drugs (Rules 90–96), the vehicle and its passengers have to fulfil certain requirements concerning roadworthiness or seatbelts (Rules 89, 97–102) and excluding exceptions listed in Rule 149, the driver cannot talk on a mobile phone in his hand. There are also some proposed changes to the Highway Code – a new section for automated vehicles is to be added. The provisions will concern driver's responsibilities and oblige the driver to be able to take back control of the vehicle, if it is designed to require the driver to resume driving after being asked to (Rules on safe use of automated vehicles on GB roads, 2022).

In 2017, the Department of Transportation in conjunction with Centre for Protection of National Infrastructure (CPNI) published the key principles of vehicle cybersecurity for connected and automated vehicles. There are eight principles, described further in sub-principles. The key principles state as follows:

1. Organizational security is owned, governed and promoted at board level;
2. Security risks are assessed and managed appropriately and proportionately, including those specific to the supply chain;

3. Organizations need product aftercare and incident response to ensure systems are secure over their lifetime;
4. Al organizations, including sub-contractors, suppliers and potential third parties, work together to enhance the security of the system;
5. Systems are designed using a defence-in-depth approach;
6. The security of all software is managed throughout its lifetime;
7. The storage and transmission of data is secure and can be controlled; and
8. The system is designed to be resilient to attacks and respond appropriately when its defences or sensors fail (HM Government, 2017, pp. 2–17).

These principles are rather general, but they convey the British Government's views on the topic and could possibly be helpful while predicting future law changes and adoption of new acts. They underline most important aspects of the law concerning automated vehicles.

In 2019, Code of Practice on automated vehicle trialling has been adopted and updated ever since. It contains guidelines on the process of trialling autonomous vehicle systems on public roads or other public places. It also provides information on driver and vehicle requirements, data recording, trials' safety, transparency and ways of cooperation with the public or authorities while planning them. Most importantly, it demands that in order to comply with the law, the organizations that perform tests "need to ensure that they have: a driver or operator, in or out of the vehicle, who is ready, able, and willing to resume control of the vehicle; a roadworthy vehicle; appropriate insurance in place".

The UK also formed a governmental Centre for Connected and Autonomous Vehicles. With its participation, in 2019, a programme named Connected and Automated Vehicles: Process for Assuring Safety and Security (CAVPASS) that involves numerous British organizations and government departments was launched. It aims to develop technical standards and regulations to ensure the safe trialling, adoption and ongoing roadworthiness of self-driving vehicles as well as processes to authorize them and to ensure that they have resilience and can respond to cyber-attacks and that their data is secure.

Its objective is also to monitor the government's actions in this field (Connected and automated vehicles: process for assuring safety and security (CAVPASS), 2022).

It is also worth underlining that in November 2023 a new government bill has been introduced into the UK Parliament. Automated Vehicles Bill is currently being worked on in the House of Lords. The final shape of the document is still unknown. The government aims to ensure clear principles of liability for the user, set the safety threshold for legal self-driving and establish a regulatory scheme to monitor the safety of these vehicles. One of the key future provisions is that an authorized self-driving vehicle will have an Authorized Self-Driving Entity assigned, that will often be a manufacturer, which will be responsible for the behaviour of the vehicle when it is self-driving (New laws to safely roll out self-driving vehicles across British roads, 2023). The bill seeks to regulate terms and processes concerning authorization of road vehicles for automated use, criminal liability for vehicle use, ways of conduct concerning policing and investigation, marketing restrictions and permits for the provision by the person of automated passenger services. The new law is to be built on the base of measures of the Automated and Electric Vehicles Act 2018 when it comes to insurance. The UK Government promises it will be "one of the world's most comprehensive legal frameworks for self-driving vehicles, with safety at its core" (Haves, 2023).

4.4.4 Singapore

Asian countries such as China, Japan or Singapore are intensely developing the autonomous vehicle technologies as well. Singapore is considered one of the world's technology leaders and is of the most frequently presented examples of innovativeness. With its modest, island territory and therefore high population density, effective and safe transportation system is of vital importance. The number of private cars on the roads is regulated, in order to diminish pollution and traffic jams. It is conducted by imposing high tax rates while purchasing a vehicle and for its everyday use. Car buyers have to pay the duty of one and a half of the car's market value, as well as through the Electronic Road Pricing (ERP) for driving in rush hours, in the city centre and for parking (Choromański et al., 2020). In order to legally register a vehicle in Singapore, a person has to bid for a Certificate of Entitlement (COE) during bidding exercises that are organized twice a month. But even after successfully securing a COE, it only enables to use the car for the next 10 years, and in order to renew a COE, it is obligatory to pay a fee called Prevailing Quota Premium (Singapore Land Transport Authority, 2023a). Because of all these factors, having a car in Singapore is formidably expensive and the city-state transport policies focus on delivering new solutions in transport, based on innovations and technologies.

In the KPMG Autonomous Vehicles Readiness Index, assessing the level of policy and legislation, technology and innovation, infrastructure and consumer acceptance, Singapore was placed at no. 1 rank in the world (KPMG International, 2020). The next 9 of 30 places in the index were allocated as follows: the Netherlands, Norway, the United States, Finland, Sweden, South Korea, United Arab Emirates, the United Kingdom and Denmark. Singapore was considered leader in both the consumer acceptance and policy and legislation. The Singapore Land Transport Authority (LTA) has been facilitating trials of autonomous vehicles since 2015. The number and area of trials expand rapidly. In 2019, the public area available to companies for trials was expanded over 1,000 km (Singapore Land Transport Authority, 2019a).

Singapore Road Traffic (Autonomous Motor Vehicles) Rules came into operation in 2017 and have been amended twice, in 2022 and 2024. Rules 4(1-2) and 7(1)(a)(i-ii) of this act state that trial or use of autonomous motor vehicles on any road is possible only if the person has a special authorization from the Singapore LTA. In granting authorization, various conditions may be imposed, as seen fit by the LTA, such as stating a geographical area in which trial or use can be conducted or prohibiting to carry passengers – Rules 9(1) and 9(2). A common requirement is that autonomous vehicles must also have a qualified safety operator or driver onboard, to take over the control of the vehicle when needed – Rule 9(2)(b-c). In order to be authorized, the vehicles must undergo stringent safety assessments conducted at the Centre of Excellence for Testing and Research of AVs-NTU (CETRAN). An enhanced testing regime for supervised trials of autonomous vehicles on public roads was implemented in February 2022. Singapore's safety assessment system for autonomous vehicles is named AV Milestone Testing Regime. It consists of three milestones, each of them including components such as document review, day and night circuit test at CETRAN, simulation test and final test on public road with the safety operator (Singapore Land Transport Authority, 2023b). When it comes to liability, the driver or operator is not responsible for the safe use of the vehicle, the liability of the driver or operator would likely only arise to the extent he is at fault under the existing road traffic rules – there is no manufacturer's liability set out, but general statutory obligations of manufacturers continue to apply (CMS, 2020). All persons authorized must place a security deposit or be equipped with comprehensive insurance coverage against third-party liability and property damage (Singapore Land Transport Authority, 2023b). Persons authorized by the LTA must ensure that the vehicle is maintained in a state of good condition and is functioning properly at all times – Rule 16(1), as well as notify the LTA about any incidents or accidents involving malfunction of the autonomous vehicle's system or death, bodily injury and damage to property – Rule 19(1). Not complying with these duties will result in the owner being guilty of an offence (CMS, 2020).

In 2019, Singapore's government published a set of provisional national standards called Technical Reference 68 (TR68), in order to guide the industry in the development and deployment of fully autonomous vehicles, while referring to SAE levels of automation (CETRAN, no date). The goal is the development of autonomous vehicle technology on SAE Levels 4 and 5. TR68 was developed by the Singapore Standards Council's Manufacturing Standards Committee, which works under the auspices of Enterprise Singapore, the national standards body. The standards contained in the document concern key spheres of the autonomous vehicle technology: vehicle basic behaviours, vehicle functional safety, cybersecurity and vehicular data types and formats. They aim to ensure safety, interoperability of data and cybersecurity (Singapore Land Transport Authority, 2019b).

Also in 2019, Singapore's Personal Data Protection Commission published the Model AI Governance Framework, providing detailed guidance to private organizations and addressing key ethical and governance issues (Personal Data Protection Commission Singapore, 2023). In 2020, it was updated with real use-cases for the first time and consideration on how AI must generate consistent results with little margin of error (Tham, 2020). But it is not the end of Singapore's AI development contribution. In 2022, the city-state's Minister for Communications and Information announced the launch of the AI Verify Foundation that is aiming to harness the contributions of the global open-source community to develop AI testing tools for the responsible use of AI (Infocom Media Development Authority, 2023).

4.5 Ethics of the AI Regulations

Parenthetically to the main theme of this chapter, it is noteworthy that some researchers bring up an interesting question, that is, "what should happen if we put an autonomous car facing the famous trolley problem" (Xiang, 2023). Although now multiple versions of the Philippa Foot's puzzle have been coined, it originally involved – of course – a trolley, that cannot be stopped, moving down the railway tracks. There are five people tied to the tracks, without any possibility to move. But it is possible to flick a switch that diverts the trolley to another track, to which a single person is tied (Tännsjö, 2013). A satisfying solution to the problem has not been reached, for there are many different views on the topic, but in a hypothetical situation, when an autonomous car would stand before such a problem, it would have to apply, in some way, an algorithm formulated by its maker. Of course, a similar problem is not likely to occur, but it is also not impossible to occur. It is also not possible to foresee every dangerous situation and prevent it from happening. In the case of car traffic, the trolley problem would probably mean a choice between harming a passerby or dodging him, simultaneously harming the driver or passengers. An exceptional problem emerges: how should the AI act? How should it be programmed? A human driver in such a situation would have to act instinctively and spontaneously, therefore in most cases, as long as he would not intentionally expand or cause damage, he would be acting unlawfully, but most likely could not be proven guilty. At the same time, self-driving car's "behaviour" would be predetermined and the accountability could be attributed to the programmers, system operators and those responsible for production and safety tests. And insurers, naturally.

In this matter, in 2017, a German Ethic Commission on Automated and Connected Driving presented its ethical rules for automated and connected vehicular traffic (Ethic Commission on Automated and

Connected Driving, 2017), where some of the answers are provided. In the fifth rule the Commission states that "the technology must be designed in such a way that critical situations do not arise in the first place" (Ethic Commission on Automated and Connected Driving, 2017). It is also clear that human life should be considered as an item of higher value than animal life or tangible property. But the Commission does not resolve the trolley problem in an unambiguous way. It states that "decisions between one human life and another depend on the actual specific situation [...] thus they cannot be clearly standardized, nor can they be programmed such that they are ethically unquestionable" (Ethic Commission on Automated and Connected Driving, 2017). The only clear statement is that "in the event of unavoidable accident situations, any distinction based on personal features [...] is strictly prohibited" and it is also prohibited to weigh lives of different people (Ethic Commission on Automated and Connected Driving, 2017).

Some point out that these rules would make the AI send a request to the driver to take over (Xiang, 2023). It can be argued that it is not an optimal solution to the problem as the driver could be unprepared and suddenly put in a difficult and stressful situation. Although other voices have appeared that show a slightly different approach. In 2020, the European Commission published Recommendations on road safety, privacy, fairness, explainability and responsibility. The 6th recommendation states that "it may be impossible to regulate the exact behaviour of CAVs [connected and automated vehicles – M.D.] in unavoidable crash situations", but it is also possible to consider the vehicles' behaviour ethical in such situations, "provided it emerges organically from a continuous statistical distribution of risk by the CAV in the pursuit of improved road safety and equality between categories of road users" (Horizon 2020 Commission Expert Group to advise on specific ethical issues raised by driverless mobility, 2020). Unfortunately, these examples show us that the AI cannot be programmed to perfectly cover all the possible future situations and provide absolute safety. Also the law cannot realistically manage autonomous vehicles in advance (Xiang, 2023). But only for now. The UNECE states that it is currently expected that a car equipped with an ADS system will have to comply with the same rules as conventional vehicles steered by drivers, for the rules demanding the driver to have the speed under control and react to changing circumstances should not be lenient for automated vehicles (United Nations Economic Commission for Europe, 2021a). The AI algorithms can and will be improved and perfected, so in the future it is possible that choosing autonomous vehicles will be the safest way of travelling by car and an optimal solution to minimize prospective harm in dangerous or unexpected situations.

Moreover, ethical issues regarding the use of AI are seen as important and they receive much attention. The High-Level Expert Group on Artificial Intelligence, set up by the European Commission, constitutes a vital part of European Union Strategy on AI and, among other acts, it published Ethic Guidelines for Trustworthy AI (European Commission, High-Level Expert Group on Artificial Intelligence, 2018). The work includes framework for achieving AI trustworthiness that is based on fundamental rights, aiming at lawful, ethical and robust AI. "Four Ethical Principles" are named there which are:

1. Respect for human authority;
2. Prevention of harm;
3. Fairness; and
4. Explicability (European Commission, High-Level Expert Group on Artificial Intelligence, 2018).

Building on these principles, "Seven Key Requirements" were adopted which are as follows:

1. Human agency and oversight;
2. Technical robustness and safety;
3. Privacy and data governance;
4. Transparency;
5. Diversity, non-discrimination and fairness;
6. Societal and environmental well-being; and
7. Accountability (European Commission, High-Level Expert Group on Artificial Intelligence, 2018).

They are to be fulfilled via technical and non-technical methods. The former include procedures implemented in the AI system's architecture, compliance with norms implemented into the design, explanation methods to better understand the system's mechanisms and find solutions, testing and validating the processing of data and introducing quality of service indicators. The latter include regulations like product safety legislation or liability frameworks, introducing codes of conduct to adopt charters of corporate policy, standardization of manufacturing and business practices, certification to attest transparency, accountability and fairness of the AI system, etc. (European Commission, High-Level Expert Group on Artificial Intelligence, 2018). Complying with these ground regulations and guidelines would certainly constitute a solid foundation for a safe and effective functioning of the AI systems that would also be acceptable for the public.

4.6 Other Smart Mobility Ideas

Connected and autonomous vehicles naturally are not the only variants of intelligent transportation based on innovative technologies. They are getting the most attention from the public and legislature, because they will probably be the most commonly known and utilized kind of transport innovation in the years to come. But there are also other ideas that could help fulfil the smart city idea and improve mass transit in future cities. Some of them are briefly described below.

4.6.1 PRT Systems

There are already examples of other means of transport operating without any human driver, for example, Personal Rapid Transit (PRT) systems, meaning electric vehicles that are able to take 2–4 people on board, operating on light, overground infrastructure (Choromański et al., 2020). These vehicles do not have a driver, they are steered with a computer. The system is designed specifically for urban areas and allows transportation on a point-to-point basis. The PRT system basically excludes the possibility of traffic jams or accidents. The current example of a functioning PRT system is Urban Light Transit (ULTra), operating on Heathrow Airport in London (Ultra Global PRT, 2011). Automated People Mover (APM) can be seen as similar to PRT systems, but the difference is that the APM moves on a railway and follows a designated route. Such a system is being used at the Phoenix Sky Harbor International Airport (Choromański et al., 2020).

4.6.2 Smart Traffic Management

Traffic management is vitally important in cities, for traffic congestion on urban road networks is increasing year by year. It causes time loss of cities' inhabitants, stress and fuel wastage. Therefore, new ways of traffic management have appeared that include ICT. Smart traffic management aims to adjust traffic lights to optimize traffic flow by detecting vehicles and pedestrians approaching an intersection. An example of this idea can be seen in Amsterdam. Virtual Traffic Manager (VTM) is an initiative providing fully automatic control of traffic (Krysiński, 2020). The project is executed by the Government of the Netherlands, the province of North Holland and the municipality of Amsterdam. VTM is connected to the governmental traffic system and therefore the systems can exchange information and react to malfunctions. The implementation of VTM in the capital of the Netherlands resulted in a 10% vehicle loss hours drop (Amsterdam Smart City, 2016). A similar project called Tristar was introduced in the Tri-City in Poland. It includes systems of traffic control, parking information, public transport management, as well as a system of signs with changing display (Krysiński, 2020).

4.6.3 Fully Automated Parking Systems

Parking is another one of the problems that contemporary cities face, because the number of personal vehicles is growing. Availability of parking space is often important when choosing a travel destination. A fully automated parking system can be one of the solutions. The parking can be placed underground or on the ground, but it needs to be multilevel to fulfil its task properly. Such systems utilize lifts to receive and return vehicles. On every level, there is a parking space and the vehicle is automatically allocated in a parking floor that is suitable for its size. The computer system ensures safe and efficient process (Opasiak and Pypno, 2021). The fully automated parking systems aim to reduce the time and space needed to park as many vehicles as possible, while remaining safe and comfortable. One of such systems was introduced in Aarhus in Denmark with space for 1,000 vehicles and retrieval time of 2–3 minutes (Dokk1, no date).

4.6.4 Street Lighting Management

Street lighting uses considerable quantities of energy. Some authorities regulate hours of its functioning. A common solution is introducing energy-efficient LED lighting. Another was designed in Barcelona, Spain. In this LED light bulbs are additionally equipped with special sensors that gather information on the environment and surroundings, especially temperature, humidity, air pollution, noise and number of pedestrians. Each factor is redirected to the central control unit, which gathers and analyses it and in effect, it can properly adjust the light's work (Krysiński, 2020, p. 168).

4.7 Conclusions

The abovementioned information constitutes only a basic fraction of our knowledge and regulations regarding autonomous vehicles and other intelligent means of transportation. Taking all of it into consideration, it can be assessed that the frequent presence of autonomous cars on our cities' streets is more of a question of "when" than a question of "if". It is happening right now. Intense research with considerable funding means fast progress. Fully self-driving cars, buses and trucks should be expected to appear more and more often in the years to come. Advanced technology, required resources and costs connected with these vehicles are the features that make them popular mainly in highly developed and urbanized areas. Other ideas of smart mobility also appear mostly in cities, for there, they are especially needed. Just like other new technologies before, their spread all across the globe is only a matter of time. Of course, further research and tests are needed to make driverless cars safer and more efficient. It should result in permission to launched this technology on our streets. The lawmakers should consider changes in the law and their possible consequences, balancing the questions of liability or required tests, but also tax rates to equalize public and private costs and benefits that are linked to this technology (Anderson et al., 2016), in order to maximize "smartness" of our cities in this sphere. For in a way, every city, big or small, is already smart.

References

Anderson, J. M. et al. (2016). Autonomous Vehicle Technology: A Guide for Policymakers, RAND Corporation, Santa Monica, CA. Available at: https://www.rand.org/pubs/research_reports/RR443-2.html (Accessed: 26.01.2024).

Blecharczyk, W. (2022). Przestrzeń SMART. Idea czy konieczność? Wydawnictwo WSEI, Kraków.

Choromański, W. et al. (2020). Pojazdy autonomiczne i systemy transportu autonomicznego, Wydawnictwo Naukowe PWN, Warszawa.

Haves, E. (2023). Automated Vehicles Bill [HL]. Available at: https://researchbriefings.files.parliament.uk/documents/LLN-2023-0045/LLN-2023-0045.pdf (Accessed: 31.01.2024).

Iwan, D. (2019). Autonomous vehicles: A new challenge to human rights?, Adam Mickiewicz University Law Review, 9. https://doi.org/10.14746/ppuam.2019.9.04

Krysiński, P. (2020). Smart city w przestrzeni informacyjnej, Wydawnictwo Naukowe Uniwersytetu Mikołaja Kopernika, Toruń.

Levin, S. and Wong, J. C. (2018). Self-driving Uber kills Arizona woman in first fatal crash involving pedestrian. Available at: https://www.theguardian.com/technology/2018/mar/19/uber-self-driving-car-kills-woman-arizona-tempe (Accessed: 13.01.2024).

Martin, T. R. (2012). Ancient Rome. From Romulus to Justinian, Yale University Press, New Haven.

Miller, C. and Valasek, C. (2015). Remote exploitation of an unaltered passenger vehicle, Proceedings of the Black Hat USA, Las Vegas, NV, USA.

Naughton, K. (2018). Just How Safe Is Driverless Car Technology, Really?, Bloomberg. Available at: https://www.bloomberg.com/news/articles/2018-03-27/just-how-safe-is-driverless-car-technology-really-quicktake (Accessed: 13.01.2024).

Olejniczak-Szuster, K. and Dziadkiewicz, M. (2020). The use of binomial models to increase digital competence of seniors in the City of Czestochowa, European Research Studies Journal, XXIII, Special Issue 1. doi: 10.35808/ersj/1789

Opasiak, T. and Pypno, C. (2021). Wielokondygnacyjny zautomatyzowany garaż nadziemny do obsługi samochodów osobowych, in Sierpiński, G. (ed.) Wyzwania stojące przed miastami przyszłości, Wydawnictwo Politechniki Śląskiej, Gliwice.

Park, S. and Choi, J.-Y. (2020). Malware detection in self-driving vehicles using machine learning Algorithms, Journal of Advanced Transportation, 2020. doi: 10.1155/2020/3035741

Rahiman, W. and Zainal, Z. (2013). An overview of development GPS navigation for autonomous car, Institute of Electrical and Electronics Engineers (IEEE) 8th Conference on Industrial Electronics and Applications (ICIEA), Melbourne Australia, VIC. doi: 10.1109/ICIEA.2013.6566533

Tännsjö, T. (2013). Understanding Ethics, 3rd edn, Edinburgh University Press, Edinburgh.

Xiang, Y. (2023). Autonomous driving ethics: Self-driving cars facing trolley problems, Lecture Notes in Education Psychology and Public Media, Vol. 15, EWA Publishing. doi: 10.54254/2753-7048/15/20231039

Legal acts:

Agreement Concerning the Adoption of Harmonized Technical United Nations Regulations for Wheeled Vehicles, equipment and parts which can be fitted and/or be used on wheeled vehicles and the conditions for reciprocal recognition of approvals granted on the basis of these United Nations Regulations, Addendum 156 – UN Regulation No. 157, Amendment 4 of 3 March 2023. Available at: https://unece.org/sites/default/files/2023-03/R157am4e%20%281%29.pdf (Accessed: 14.01.2024).

Agreement concerning the adoption of harmonized technical United Nations Regulations for wheeled vehicles, equipment and parts which can be fitted and/or be used on wheeled vehicles and the conditions for reciprocal recognition of approvals granted on the basis of these United Nations Regulations of 20 March 1958. Available at: https://treaties.un.org/Pages/ViewDetails.aspx?src=IND&mtdsg_no=XI-B-16&chapter=11&clang=_en (Accessed: 12.01.2024).

Agreement concerning the Establishing of Global Technical Regulations for Wheeled Vehicles, Equipment and Parts which can be fitted and/or be used on Wheeled Vehicles of 25 June 1998. Available at: https://treaties.un.org/Pages/ViewDetails.aspx?src=TREATY&mtdsg_no=XI-B-32&chapter=11&clang=_en (Accessed: 12.01.2024).

Automated and Electric Vehicles Act. (2018). Available at: https://www.legislation.gov.uk/ukpga/2018/18/contents/enacted (Accessed: 26.01.2024).

Automated Vehicles Bill. (2023). HL Bill no. 1, 2023-2024. Available at: https://lordslibrary.parliament.uk/research-briefings/lln-2023-0045/ (Accessed: 30.01.2024).

California Vehicle Code, CA Veh Code § 38750. Available at: https://law.justia.com/codes/california/2012/veh/division-16.6/section-38750 (Accessed: 27.01.2024).

Code of Practice: Automated vehicle trialling. (2019). Available at: https://www.gov.uk/government/publications/trialling-automated-vehicle-technologies-in-public/code-of-practice-automated-vehicle-trialling (Accessed: 30.01.2024).

Commission Delegated Regulation (EU) 2022/2236 of 20 June 2022 amending Annexes I, II, IV and V to Regulation (EU) 2018/858 of 20 June 2022 of the European Parliament and of the Council as regards the technical requirements for vehicles produced in unlimited series, vehicles produced in small series, fully automated vehicles produced in small series and special purpose vehicles, and as regards software update. Available at: https://eur-lex.europa.eu/eli/reg_del/2022/2236/oj (Accessed: 01.02.2024).

Commission Implementing Regulation (EU) 2022/1426 of 5 August 2022 laying down rules for the application of Regulation (EU) 2019/2144 of the European Parliament and of the Council as regards uniform procedures and technical specifications for the type-approval of the automated driving system (ADS) of fully automated vehicles. Available at: https://eur-lex.europa.eu/legal-content/EN/TXT/?uri=CELEX%3A32022R1426 (Accessed: 01.02.2024).

Communication from the Commission of 15 May 2018. On the road to automated mobility: An EU strategy for mobility of the future. Available at: https://eur-lex.europa.eu/legal-content/EN/TXT/?uri=CELEX%3A52018DC0283 (Accessed: 01.02.2024).

Convention on Road Traffic of 8 November 1968. Available at: https://treaties.un.org/Pages/ViewDetailsIII.aspx?src=TREATY&mtdsg_no=XI-B-19&chapter=11&Temp=mtdsg3&clang=_en (Accessed: 12.01.2024).

Declaration of Amsterdam on cooperation in the field of connected and automated driving 2016. Available at: https://bmdv.bund.de/SharedDocs/DE/Anlage/DG/amsterdamer-erklaerung-declaration-of-amsterdam.pdf?__blob=publicationFile (Accessed: 01.02.2024).

Federal Motor Vehicle Safety Standards, 49 CFR 571.101-500. Available at: https://www.ecfr.gov/current/title-49/subtitle-B/chapter-V/part-571 (Accessed: 26.01.2024).

Framework for Automated Driving System Safety. (2020). 85 FR 78058. Available at: https://www.federalregister.gov/documents/2020/12/03/2020-25930/framework-for-automated-driving-system-safety (Accessed: 26.01.2024).

Occupant Protection for Automated Driving Systems. (2020). 85 FR 17624. Available at: https://www.federalregister.gov/documents/2020/03/30/2020-05886/occupant-protection-for-automated-driving-systems (Accessed: 26.01.2024).

Occupant Protection for Vehicles with Automated Driving Systems. (2022). 87 FR 18560. Available at: https://www.federalregister.gov/documents/2022/03/30/2022-05426/occupant-protection-for-vehicles-with-automated-driving-systems (Accessed: 26.01.2024).

Regulation (EU) 2018/858 of the European Parliament and of the Council of 30 May 2018 on the approval and market surveillance of motor vehicles and their trailers, and of systems, components and separate technical units intended for such vehicles. Available at: https://eur-lex.europa.eu/legal-content/EN/ALL/?uri=celex:32018R0858 (Accessed: 01.02.2024).

Regulation (EU) 2019/2144 of the European Parliament and of the Council of 27 November 2019 on type-approval requirements for motor vehicles and their trailers, and systems, components and separate technical units intended for such vehicles, as regards their general safety and the protection of vehicle occupants and vulnerable road users. Available at: https://eur-lex.europa.eu/legal-content/EN/ALL/?uri=CELEX:32019R2144 (Accessed: 01.02.2024).

Road Traffic (Autonomous Motor Vehicles) Rules 2017, https://sso.agc.gov.sg/SL/RTA1961-S464-2017?DocDate=20240102&ValidDate=20240101&Timeline=On&ProvIds=P11-#pr1- (Accessed: 31.01.2024).

The Highway Code. Available at: https://www.highwaycodeuk.co.uk (Accessed: 26.01.2024).

U.S. Department of Transportation, National Highway Traffic Safety Administration (2023) Second Amended Standing General Order 2021-01. Available at: https://www.nhtsa.gov/sites/nhtsa.gov/files/2023-04/Second-Amended-SGO-2021-01_2023-04-05_2.pdf (Accessed: 27.01.2024).

Internet sources:

Amsterdam Smart City. (2016). Smart traffic management. Available at: https://amsterdamsmartcity.com/updates/project/smart-traffic-management (Accessed: 02.02.2024).

California Department of Motor Vehicles. (2024). Autonomous Vehicle Collision Reports. Available at: https://www.dmv.ca.gov/portal/vehicle-industry-services/autonomous-vehicles/autonomous-vehicle-collision-reports/ (Accessed: 02.02.2024).

CETRAN. (no date). Singapore Publishes Technical Reference for AVs. Available at: https://onemotoring.lta.gov.sg/content/onemotoring/home/buying/upfront-vehicle-costs/certificate-of-entitlement--coe-.html (Accessed: 12.01.2024).

CMS. (2020). Autonomous Vehicles Law and Regulation in Singapore. Available at: https://cms.law/en/int/expert-guides/cms-expert-guide-to-autonomous-vehicles-avs/singapore (Accessed: 14.01.2024).

Connected and automated vehicles: process for assuring safety and security (CAVPASS). (2022). Available at: https://www.gov.uk/guidance/connected-and-automated-vehicles-process-for-assuring-safety-and-security-cavpass (Accessed: 29.02.2024).

Dokk1. (no date). How it works. Available at: https://dokk1-parkering.dk/en/the-car-park.aspx (Accessed: 02.02.2024).

Ethic Commission on Automated and Connected Driving (2017) Report. Available at: https://bmdv.bund.de/SharedDocs/EN/publications/report-ethics-commission.pdf?__blob=publicationFile (Accessed: 26.01.2024).

European Commission, High-Level Expert Group on Artificial Intelligence (2018) Draft Ethics Guidelines for Trustworthy AI. Available at: https://digital-strategy.ec.europa.eu/en/policies/expert-group-ai (Accessed: 26.01.2024).

European Commission. (2022). New rules to improve road safety and enable fully driverless vehicles in the EU. Available at: https://ec.europa.eu/commission/presscorner/detail/en/ip_22_4312 (Accessed: 01.02.2024).

FRAV, Informal Working Group on Functional Requirements for Automated Vehicles (2020) Report to the 7th GRVA Session. Available at: https://unece.org/fileadmin/DAM/trans/doc/2020/wp29grva/GRVA-07-54e.pdf (Accessed: 14.01.2024).

GRVA (2022) New Assessment/Test Method for Automated Driving (NATM) Guidelines for Validating Automated Driving System (ADS) – amendments to ECE/TRANS/WP.29/2022/58. Available at: https://unece.org/sites/default/files/2022-05/WP.29-187-08e.pdf (Accessed: 14.01.2024).

High Level Group on the Competitiveness and Sustainable Growth of the Automotive Industry in the European Union (GEAR 2030) (2017)

Ensuring that Europe has the most competitive, innovative and sustainable automotive industry of the 2030s and beyond. Available at: https://www.europarl.europa.eu/cmsdata/141562/GEAR%202030%20Final%20Report.pdf (Accessed: 01.02.2024).

HM Government. (2017). The key principles of cyber security for connected and automated vehicles. Available at: https://assets.publishing.service.gov.uk/media/5a8229c2ed915d74e62361f1/cyber-security-connected-automated-vehicles-key-principles.pdf (Accessed: 29.01.2024).

Horizon 2020 Commission Expert Group to advise on specific ethical issues raised by driverless mobility. (2020) Ethics of Connected and Automated Vehicles: Recommendations on Road Safety, Privacy, Fairness, Explainability and Responsibility, Publication Office of the European Union, Luxembourg. Available at: https://op.europa.eu/en/publication-detail/-/publication/89624e2c-f98c-11ea-b44f-01aa75ed71a1/language-en (Accessed: 02.02.2024).

Infocom Media Development Authority. (2023). Singapore launches AI Verify Foundation to shape the future of international AI standards through collaboration. Available at: https://www.imda.gov.sg/resources/press-releases-factsheets-and-speeches/press-releases/2023/singapore-launches-ai-verify-foundation-to-shape-the-future-of-international-ai-standards-through-collaboration (Accessed: 30.01.2024).

Jones Day. (2021). White Paper. Autonomous Vehicles: Legal and Regulatory Developments in the United States. Available at: https://www.jonesday.com/en/insights/2021/05/autonomous-vehicles-legal-and-regulatory-developments-in-the-us (Accessed: 26.01.2024).

KPMG International. (2020). 2020 Autonomous Vehicles Readiness Index. Assessing the preparedness of 30 countries and jurisdictions in the race for autonomous vehicles. Available at: https://assets.kpmg.com/content/dam/kpmg/xx/pdf/2020/07/2020-autonomous-vehicles-readiness-index.pdf (Accessed: 15.01.2024).

New laws to safely roll out self-driving vehicles across British roads (2023). Available at: https://www.gov.uk/government/news/new-laws-to-safely-roll-out-self-driving-vehicles-across-british-roads (Accessed: 30.01.2024).

Personal Data Protection Commission Singapore. (2023). Singapore's Approach to AI Governance. Available at: https://www.pdpc.gov.sg/Help-and-Resources/2020/01/Model-AI-Governance-Framework (Accessed: 30.01.2024).

Quidachay-Swan, S. (2019). Autonomous vehicles and current state liability Legislation, Michigan Bar Journal, 98 (3). Available at: http://www.michbar.org/file/barjournal/article/documents/pdf4article3624.pdf (Accessed: 27.01.2024).

Riess, R. and Sottile, Z. (2023). Uber self-driving car test driver pleads guilty to endangerment in pedestrian death case. Available at: https://edition.cnn.com/2023/07/29/business/uber-self-driving-car-death-guilty/index.html (Accessed: 13.01.2024).

Rules on safe use of automated vehicles on GB roads (2022). Available at: https://www.gov.uk/government/consultations/safe-use-rules-for-automated-vehicles-av/rules-on-safe-use-of-automated-vehicles-on-gb-roads (Accessed: 31.01.2024).

San Francisco Municipal Transportation Agency. (2019). Driving Automation Systems: Advanced Driver Assistance Systems (ADAS) and Automated Driving Systems (ADS). Available at: https://www.sfmta.com/projects/driving-automation-systems-advanced-driver-assistance-systems-adas-and-automated-driving (Accessed: 13.01.2024).

SAE International. (2018). SAE International Releases Updated Visual Chart for Its "Levels of Driving Automation" Standard for Self-Driving Vehicles. Available at: https://www.sae.org/news/press-room/2018/12/sae-international-releases-updated-visual-chart-for-its-%E2%80%9Clevels-of-driving-automation%E2%80%9D-standard-for-self-driving-vehicles (Accessed: 15.01.2024).

SAE International. (2021). SAE Levels of Driving Automation™ Refined for Clarity and International Audience. Available at: https://www.sae.org/blog/sae-j3016-update (Accessed: 24.01.2024).

Self-driving vehicles listed for use in Great Britain. (2022). Available at: https://www.gov.uk/guidance/self-driving-vehicles-listed-for-use-in-great-britain (Accessed: 29.01.2024).

Singapore Land Transport Authority. (2019a). Autonomous Vehicle Testbed to be Expanded to Western Singapore – Continued Emphasis on Public Safety. Available at: https://www.lta.gov.sg/content/ltagov/en/newsroom/2019/10/1/Autonomous_vehicle_testbed_to_be_expanded.html#2 (Accessed: 14.01.2024).

Singapore Land Transport Authority. (2019b). Joint Media Release by the Land Transport Authority (LTA), Enterprise Singapore, Standards Development Organisation & Singapore Standards Council – Singapore Develops Provisional National Standards to Guide Development of Fully Autonomous Vehicles. Available at: https://www.lta.gov.sg/content/ltagov/en/newsroom/2019/1/2/joint-media-release-by-the-land-transport-authority-lta-enterprise-singapore-standards-development-organisation-singapo.html#2 (Accessed: 14.01.2024).

Singapore Land Transport Authority. (2023a). Certificate of Entitlement (COE).

Singapore Land Transport Authority. (2023b). Autonomous Vehicles. Available at: https://www.lta.gov.sg/content/ltagov/en/industry_innovations/technologies/autonomous_vehicles.html (Accessed: 14.01.2024).

Statistisches Bundesamt. (2023). The largest cities worldwide 2023. Available at: https://www.destatis.de/EN/Themes/Countries-Regions/International-Statistics/Data-Topic/Population-Labour-Social-Issues/Demography Migration/UrbanPopulation.html (Accessed: 10.01.2024).

Steiger, R. W. (1995). Roads of the Roman Empire. Available at: https://pita.ess.washington.edu/tswanson/wp-content/uploads/sites/9/2018/10/Roads-of-the-Roman-Empire.pdf (Accessed: 05.01.2024).

Tencent Keen Security Lab. (2019). Tencent Keen Security Lab: Experimental Security Research of Tesla Autopilot. Available at: https://keenlab.tencent.com/en/2019/03/29/Tencent-Keen-Security-Lab-Experimental-Security-Research-of-Tesla-Autopilot/ (Accessed: 13.01.2024).

Tham, I. (2020). World Economic Forum: Singapore updates AI governance model with real-world cases. Available at: https://www.straitstimes.com/world/spore-updates-ai-governance-model-with-real-world-cases (Accessed: 30.01.2024).

Ultra Global PRT. (2011). Heathrow pods transport passengers to the future. Available at: https://www.ultraglobalprt.com/heathrow-pods-transport-passengers-to-the-future/ (Accessed: 17.01.2024).

United Nations Economic Commission for Europe. (2022). UN Regulation extends automated driving up to 130 km/h in certain conditions. Available at: https://unece.org/sustainable-development/press/un-regulation-extends-automated-driving-130-kmh-certain-conditions (Accessed: 13.01.2024).

United Nations Economic Commission for Europe. (2021a). All you need to know about Automated Vehicles. Technical progress and regulatory activities. Available at: https://unece.org/sites/default/files/2022-01/Brochure%20Automated%20Vehicles.pdf (Accessed: 13.01.2024).

United Nations Economic Commission for Europe. (2021b). UN regulation on Automated Lane Keeping Systems (ALKS) extended to trucks, buses and coaches. Available at: https://unece.org/media/transport/Vehicle-Regulations/press/362551 (Accessed: 13.01.2024).

United Nations, Department of Economic and Social Affairs, Population Division. (2018). World Urbanization Prospects: The 2018 Revision. File 2: Percentage of Population at Mid-Year Residing in Urban Areas by Region, Subregion, Country and Area, 1950-2050. Available at: https://population.un.org/wup/Download/ (Accessed: 09.01.2024).

UNCTAD. (2023). Total and urban population. Available at: https://hbs.unctad.org/total-and-urban-population/ (Accessed: 09.01.2024).

U.S. Department of Transportation. (2021). Ensuring American Leadership in Automated Vehicle Technologies: Automated Vehicles 4.0. Available at: https://www.transportation.gov/av/4 (Accessed: 26.01.2024).

U.S. Department of Transportation, National Highway Traffic Safety Administration. (2017). Automated Driving Systems: A Vision for Safety 2.0. Available at: https://www.nhtsa.gov/sites/nhtsa.gov/files/documents/13069a-ads2.0_090617_v9a_tag.pdf (Accessed: 26.01.2024).

U.S. Department of Transportation, National Science and Technology Council. (2020). Ensuring American Leadership in Automated Vehicle Technologies: Automated Vehicles 4.0. Available at: https://www.transportation.gov/sites/dot.gov/files/2020-02/EnsuringAmericanLeadershipAVTech4.pdf (Accessed: 26.01.2024).

U.S. Department of Transportation, National Highway Traffic Safety Administration. (2016). Federal Automated Vehicles Policy. Accelerating the Next Revolution in Roadway Safety. Available at: https://www.transportation.gov/sites/dot.gov/files/docs/AV%20policy%20guidance%20PDF.pdf (Accessed: 26.01.2024).

U.S. Department of Transportation, National Highway Traffic Safety Administration. (2023a). Traffic Safety Facts. 2021 Data. Available at: https://crashstats.nhtsa.dot.gov/Api/Public/ViewPublication/813488 (Accessed: 09.01.2024).

U.S. Department of Transportation, National Highway Traffic Safety Administration. (2023b). Standing General Order on Crash Reporting For incidents involving ADS and Level 2 ADAS. Available at: https://www.nhtsa.gov/laws-regulations/standing-general-order-crash-reporting (Accessed: 13.01.2024).

U.S. Department of Transportation. (2018). Preparing for the Future of Transportation: Automated Vehicles 3.0. Available at: https://www.transportation.gov/sites/dot.gov/files/docs/policy-initiatives/automated-vehicles/320711/preparing-future-transportation-automated-vehicle-30.pdf (Accessed: 26.01.2024).

World Health Organization. (2023). Road Traffic Injuries. Available at: https://www.who.int/news-room/fact-sheets/detail/road-traffic-injuries (Accessed: 10.01.2024).

5

Tax Incentives for Socially Responsible Companies

IZABELA BAGIŃSKA

5.1 Introduction

Although the origins of social responsibility can be traced back to antiquity, the idea of corporate social responsibility (CSR), which is still valid today, originated in the 20th century. It was in the last century that the concept of sustainable development began to develop, which was visible in the growing interest of entrepreneurs in charitable activities or involvement in environmental issues. Undoubtedly, the environmental disasters that took place in the 1970s and 1980s also led to an increase in social and environmental awareness.

Leo XII and his 1891 encyclical *Rerum Novarum* is considered the forerunner to social responsibility, while on American soil it was the publication of E. W. Lord's book *The Fundamentals of Business Ethics* in 1926 (Bernatt, 2009) that led the way. Then, in 1932, Merrick Dodd proclaimed that the goal of an entrepreneur should not only be to make a profit, but also to act in the social interest. Since entrepreneurs belong to a certain social environment, they therefore have certain rights and obligations to fulfil for the benefit of this environment (Dodd, 1932).

Another author who linked businesses to the social environment was the American economist Howard Bowen. In his work *The Social Responsibility of the Businessman*, it was entrepreneurs who he made responsible for the condition of society (Bowen, 1953). He was also the first to use the phrase: "social responsibility".

Climate change and increasing environmental degradation are negatively affecting the sustainability of business areas around the world. Along with other driving factors, such as globalization and rapid technological change, they are causing significant changes in labor markets.

DOI: 10.1201/9781003465157-5

The development of globalization and increased public awareness have made entrepreneurs more willing to get involved in social and environmental issues. As a result, they were able to improve their image in the market and strengthen their credibility among their customers (Jha and Cox, 2015). Moreover, socially responsible activities have a positive impact on society and the environment by, for example, creating new workplaces, reducing greenhouse gas emissions, reducing energy or water consumption.

The increased interest of companies in implementing CSR stems largely from the promotion of CSR by the socio-economic environment. Factors that foster the implementation of CSR into the activities of enterprises include the ISO 26000 Standard, the SA8000 Standard, the AA1000 Series Standard, the appointment of the CSR Team by the Minister of Economy, the inclusion of CSR in strategic documents together with the Enterprise Development Program until 2020 (Bereś, 2017).

In 2000, during the European Union (EU) summit, the European Council adopted the Lisbon Strategy, which was an expression of interest in CSR. Its goal was to improve the knowledge-based economy capable of creating new workplaces over the next 10 years. The participation of entrepreneurs, to whom a special appeal was made for the use of socially responsible activities, was recognized as a prerequisite for achieving the set goal. Since the Lisbon summit, the issue of CSR has become a priority topic in the EU (Jastrzębska, 2011).

The sustainability agenda is complex and requires a variety of actions that establish principles and goals that affect the actions of individuals and society at large. The effects of these activities can have serious implications for the skills programme, which must evolve in response to these changes (Oana and Dina i Catalin Martin, 2011).

The concept of sustainability is reflected in the policies of individual EU countries. Although the CSR concept takes into account voluntary actions of companies for the benefit of society and the environment, public institutions in some countries, recognizing the benefits of its application, encourage companies by offering certain forms of support (https://www.cire.pl/pliki/2/12022Raport.pdf).

It is not only the companies that have an interest in CSR. The state also has an interest in promoting this way of management. One of the reasons may certainly be the recommendations of international organizations concerning the implementation of the discussed concept in business entities, along with trends emerging in the economy and society, related to the idea of social responsibility.

5.2 What Is Corporate Social Responsibility?

One of the definitions of CSR states that it is a concept whereby companies voluntarily integrate social, ethical and environmental aspects into their operations (Commission of the European Communities). Socially responsible activities revolve around human resources, environmental protection, ethical actions in relations with employees.

Some authors describe CSR as a set of activities aimed at promoting some social good which goes beyond what is part of an organization's ordinary activities (Babiak and Trendafilova, 2011).

As the concept of social responsibility is gaining interest worldwide, it has also been recognized by the International Organization for Standardization (ISO), which, in 2005, appointed the *ISO Working Group on Social Responsibility*. It consisted of some 450 experts and 210 observers from 99 ISO member countries and 42 affiliated organizations. On November 1, 2010, the *ISO 26000 Guidance on Social Responsibility* was issued which is considered one of the most important ISO standards (Rojek-Nowosielska, 2017).

It seems true to say that in order to improve existing practices and further integrate social responsibility into organizations, it is necessary to use the ISO 26000 standard (Balzarova and Castka, 2012). It is worth mentioning that the aforementioned standard is applicable to all organizations, whether public, private or non-profit, regardless of their size. Therefore, it does not apply merely to CSR.

According to the definition proposed by the ISO 26000, social responsibility should be understood as an organization's responsibility for the impact of its decisions and actions on society and the environment, through transparent and ethical conduct which (Rojek-Nowosielska, 2017):

1. Contributes to sustainable development, including the health and well-being of society,
2. Takes into account stakeholders' expectations,

3. Is consistent with applicable laws and in compliance with international standards of conduct,
4. Is consistent with the organization's activities and practised in its relationships, which pertain to the organization's activities undertaken within its sphere of influence.

The implementation of the ISO 26000 guidelines can positively affect competitiveness, reputation, ability to attract and retain employees, investor perceptions, relations with businesses, governments, suppliers or partners (Fujarska and Szewczyk, 2011).

The guidelines described in ISO 26000 concern seven subject areas. The first of these addresses organizational governance. Good practices related to this area aim to improve the efficiency of business management, taking into account the public interest, respect for stakeholders and ethical principles.

The second area is related to human rights with two categories. The first category pertains to civil and political rights and includes the right to life and liberty or the right to equal treatment. The second category relates to economic, social and cultural rights, and therefore includes the right to work, the right to food and the right to education. Good practices in this area revolve around education or equal opportunities for women.

The third area mentioned is related to labor issues. A company's activities in this area go beyond the relationship with its direct employees or the organization's responsibilities in relation to the workplace it has. Good practices in this area should take care of working conditions, occupational health and safety, social development (training), as well as bearing in mind the need for ongoing social dialogue and maintaining open and honest relations with cooperating entities.

Another area is related to the environment. It is recognized that the activities of any organization always affect the environment, therefore it is recommended to adopt an integrated approach which takes into account both direct and indirect economic, social, health and environmental effects on its activities. Good practices should focus on monitoring and reducing pollution emitted into the environment, as well as taking measures to reduce the level of natural resources consumption.

The fifth area relates to business ethics and is particularly concerned with the ethical behavior of the company in its relations with other entities. Good practices in this area aim to counteract unfair competition, embezzlement and corruption, and help promote fair cooperation, social responsibility in the supply chain and respect for property rights.

Another area addresses consumer issues. Any company that provides certain goods or services to a customer has certain obligations toward that customer. First of all, it should take care of proper marketing, present clear and precise information so as not to mislead the customers. Fair sales rules should apply, as well as consumer safety should be in focus.

The last area pertains to the social development of the company which should engage in social activities. Good practices in this area relate to the ability to conduct social dialogue or implement social projects. Companies should undertake activities in the areas of health, education, culture or access to technology.

As defined in the Green Paper, CSR means voluntarily integrating social and environmental aspects into one's strategy, both in commercial activities and in dealing with other stakeholders. The goal of CSR is to take responsibility for society and the environment (Green Paper, 2001).

The issue of social responsibility was also noted at the World Economic Forum in Davos in 1999 by the creators of the Global Compact idea. The authors urge entrepreneurs to follow certain principles in their business activities from the field of CSR (Corporate Social Responsibility: Report, 2008):

1. Human rights,
2. Labor standards,
3. Environmental protection,
4. Anti-corruption.

Archie Carroll introduced four types of business responsibilities: economic, legal, ethical and philanthropic. He further pointed out that they are at different levels, and for this reason a company should approach them in the right order (Carroll, 1991).

According to the author of this concept, the key area is economic responsibility. The company is oriented at maximizing profit, minimizing costs, improving efficiency and strengthening its competitive position. It seems necessary for the company to take responsibility for its own actions in this regard.

The next type of responsibility required is legal responsibility, that is, actions in compliance with laws, environmental protection or consumer rights.

At the next level of the pyramid is ethical responsibility which embraces all honest and fair corporate actions.

And at the top of the pyramid is philanthropic responsibility, the most desired by society. It includes any education, health or local community activities.

The main areas of social responsibility are indicated by D. Aarts. According to him, these are the workplace, the environment, the community and the market (Aarts, 2011).

An interesting perspective on the issue of social responsibility is presented by M. Porter and M. Kramer, as they stated that there are mutual interactions between the enterprise and the environment (Porter and Kramer, 2011). They pointed out that the enterprise creates new jobs for society, while the environment provides certain resources. Two types of activities can be distinguished in this connection:

1. The actions taken by the enterprise and the way they are perceived by the environment.
2. Social activities and their impact on the enterprise.

The authors of this concept take the stand that the proper use of both types of activities ensures the achievement of shared value.

Nowadays, social and environmental activities are expected to help complement the socially responsible expectations placed on modern organizations seeking sustainable development in a competitive, rapidly changing market.

Socially responsible companies make a special effort to include their concern for other stakeholders in their operations (Carroll, 2015)

5.3 Selected Tax Instruments to Support CSR

CSR is an important element of the EU's strategy. This is confirmed, among other things, by the strategy "Europe 2020 - Strategy for Smart, Sustainable and Inclusive Growth" and the Communication from the Commission to the European Parliament, the Council, the European Economic and Social Committee and the Committee of the Regions,

"Renewed European Union Strategy 2011–2014 for Corporate Social Responsibility" (Leoński, 2016).

Pursuant to the Order of the Minister of Funds and Regional Policy dated January 21, 2020, a Team for Sustainable Development and Corporate Social Responsibility was established. Its tasks include (https://www.gov.pl/web/fundusze-regiony/o-zespole):

1. Conducting dialogue, exchange of experiences and good practices between public administration, business, socio-economic partners in the field of sustainable development and CSR,
2. Disseminating the principle of social solidarity and responsible business conduct,
3. Disseminating the universal idea of social responsibility of organizations in the environment of universities and public administration.

When considering the role of the state in the development of CSR, the first thing that should be taken into account is the efforts to modify laws in the direction of promoting pro-environmental and pro-social attitudes among entrepreneurs. We are talking about the changes in the tax law.

Companies tend to be reluctant to pay taxes, which results in aggressive tax planning. This is connected with the phenomenon of tax avoidance, which occurs in different intensities and forms in individual countries, as well as with tax evasion.

CSR should be considered from the perspective of tax planning. It is a common rule of thumb that businesses, in order to reduce their obligations to the tax authorities, use various tools offered by the state. This in turn often leads to aggressive tax planning. One should agree with the statement that the line between legal tax avoidance and illegal tax evasion is blurred (Duff, 2009).

In an attempt to clarify what tax avoidance consists of and whether it is an illegal phenomenon from the point of view of the law, it is necessary to refer to the Tax Ordinance. According to it, tax avoidance is understood as actions performed for the purpose of obtaining a tax benefit that, under the given circumstances, is contrary to the object and purpose of the provision of the law. If the tax authorities prove that the taxpayer's course of action was artificial, the act does not result in gaining an advantage (Tax Ordinance, article 119a, paragraph 1).

The characteristic features of tax avoidance include the following:

1. It is in accordance with the letter, but contrary to the spirit of the tax law (ratio legis);
2. It leads to the taxpayer's tax benefit in the form of reduction or elimination of tax liabilities,
3. The motive for a particular action is to achieve a tax benefit.

In the subject literature, one can encounter the approach that due to the fact that tax avoidance consists in reducing the tax burden by methods and means permitted by the tax law, it is not an unethical activity (Gomułovich and Małecki, 2013). However, considering this problem from the point of view of ethics, it would be necessary to consider such an action as unethical.

Tax avoidance is contrasted with tax evasion and relies on the failure to reveal a tax obligation. Compared to tax avoidance, tax evasion is definitely an illegal action.

Aggressive tax planning by companies is a common phenomenon. It is difficult to deal with it, despite the implementation of more and more new instruments that limit the taxpayers' freedom in shaping their tax bases. There is a view in the subject literature that entrepreneurs should focus on maximizing corporate profits, not taxes. Taxes should remain the concern of individual states (Stephenson and Vracheva, 2016).

Changes in laws and regulations are necessary, but they will not solve all the problems associated with tax avoidance. Increased public interest and public pressure have also led some companies to view paying taxes to the budget as part of a policy related to running a sustainable business and CSR (Wasilewski and Bischof, 2017).

Environmental or social activities involve incurring various types of costs, which is why tax incentives offered by the state are becoming more common in the context of CSR.

By reducing the tax burden, they foster environmental, social and economic development. By offering various concessions to socially responsible entities, the state influences the behaviour of taxpayers, either encouraging them to behave in a certain way or, on the contrary, discouraging them from taking certain actions (Kacem and Omri, 2022).

Based on analysed studies and reports on CSR issues, the predominant types of so-called good practices of sustainable business in the tax context were identified.

5.3.1 Special Economic Zones

In 1994, the Law of October 20, 1994 on special economic zones (SEZs) came into force. According to it, SEZs are an uninhabited part of the territory of the Republic of Poland, separated in accordance with the provisions of the Act, on the territory of which economic activity may be carried out under the rules set forth in the Act. The most commonly used tax policy instruments are exemptions, lower tax rates and tax concessions. One of the key issues in the functioning of SEZs is the effectiveness of the tax concessions applied in a given zone (Sinenko, 2016).

Since the establishment of SEZ, the idea of their operation has remained virtually unchanged. It is a geographically defined area that is supposed to bring a number of benefits to entities. They were created to accelerate the economic development of underdeveloped regions. Table 5.1 shows the tax incentives for activities carried out in SEZs (Accreo Taxand Ltd.) in selected countries.

Currently, there is no tax exemption available under SEZs and they are only scheduled to operate until the end of 2026.

The development of science and technology, which has been observed for many years, has led to environmental degradation. The aim of the international effort was to identify the most important global threats and develop ways to reduce and counter them. Among the most important problems were (Gołąbewska and Harasimowicz, 2023):

1. Threats to the earth's atmosphere,
2. Degradation of the earth's surface,
3. Water shortage,
4. Threats to species diversity,
5. Destruction of forests, and
6. Famine.

There are various categories of incentives provided by the authorities of individual countries to promote green building. Among them, the most noteworthy are tax incentives concerning the property tax establishment rules. They have been applied in the legislation of Spain, Romania, Italy, Bulgaria, the United States, Canada, Malaysia and India, to name a few (Shazmin, Sipan and Sapri, 2016).

It is important to agree with the statement that the construction sector is responsible to a significant extent for the environmental

Table 5.1 Tax Incentives for SEZs in Selected Countries.

Poland	1. Under certain conditions, income from business conducted within the SEZ on the basis is subject to exemption from the income tax (CIT, PIT). 2. Exemptions from the immovable property tax, which applies to land, buildings or their parts as well as structures or their parts related to the running of business activities. 3. In order to encourage non-agricultural work, tax preferences were introduced in the Law on Lump-Sum Income Tax on Certain Income Earned by Natural Persons for taxpayers running farms, who at the same time carry out non-agricultural economic activity, in towns with a population of up to 5,000.
France	1. The exemption from taxation or reduction in the rate of CIT and local business tax for companies operating in less developed regions. These reliefs apply mainly to the French islands (e.g., Corsica), overseas territories and rural areas.
Norway	1. Geographical differentiation of the amount of mandatory social security contributions paid by the employer. The main purpose of such an arrangement is to stimulate business development, prevent the liquidation of enterprises, as well as provide sustainable employment for the residents of developing regions. A lower tax rate is applied to the income of both individual and corporate investors who reside on Svalbard. 2. Tax relief in the form of accelerated depreciation to oil production platforms outside the Finnmark administrative district.
Finland	1. Accelerated depreciation in developing countries.
Greece	1. Tax-free amounts, depending on the region. 2. Reduced VAT rate for the supply of goods and services on certain islands.
Germany	1. Income tax relief or non-refundable grant.
Romania	1. Exemption from the construction tax and the immovable property tax. 2. Exemption from the income tax in connection with the implementation of new investments.
Switzerland	1. Income tax reliefs.

Source: Review of tax incentives in selected European countries, ACCreo Taxand, Warsaw 2011; Special Economic Zones after 2020. Analysis of current activities and prospects for operation, Ernst & Young, 2011.

pollution, including high levels of energy consumption, greenhouse gas emissions and resource depletion (Mekhilef, Saidur and Safar, 2011). Studies conducted on the impact of tax incentives on the development of green building indicate that property tax credits have the greatest impact in this regard (Cansino, 2011; Clement and Lehman, 2005)

The state offers entrepreneurs a package of reliefs they can use when engaging in socially beneficial activities. When evaluating the state's efforts to promote CSR in the area of tax law, it is worth noting that new reliefs have recently been introduced to encourage entrepreneurs to engage in socially responsible activities. Cooperation between the business sector and national governments seems essential.

Countries applying the Research and Development Tax Credit include:

1. Spain introduced a tax credit consisting of a 25% tax deduction for expenses incurred on research and development (R&D) activities in a given tax year.
2. France, where an R&D tax credit has been introduced for investment and implementation of environmental innovations at 30% of expenses incurred by existing companies and 35–40% for new entities. These entities can deduct the expenses from the income tax in the following three tax years.
3. Greece, where there is an allowance dedicated to manufacturing facilities that incur costs to reduce the environmental impact of their operations. They can reduce their tax base by 50% of the value of such expenses.
4. Belgium, which has introduced a tax credit of 80% of the expenditures, invested to acquire a patent under certain conditions.
5. Romania, which has introduced a research and development tax credit that provides an additional tax deduction of 20% of the value of eligible expenses and the use of an accelerated depreciation method for devices and equipment intended for research and development activities.
6. Poland, where taxpayers who carry out research and development activities and are subject to progressive or flat tax on these activities are entitled to a research and development (R&D) tax credit. The relief consists of deducting from the tax base a part of the tax-deductible expenses incurred for these activities.

Eligible deductible expenses include, for example (Corporate Income Tax Law, Article 18d):

1. Salaries of employees in the part related to R&D activities, as well as related social contributions;
2. Remuneration for civil law contracts in the part related to R&D activity, as well as social contributions related to them;
3. Expenditures on the acquisition of specialized equipment and materials and raw materials that are directly related to R&D activity;

4. Expenditures on expert opinions, opinions, consulting services and equivalent services provided or performed under contract by entities referred to in the regulations on higher education and science, as well as on the acquisition of results of scientific research conducted by them, for the purposes of R&D activity;
5. Expenditures on the acquisition or use, against payment, of scientific and research apparatus used exclusively in R&D activities;
6. Costs of obtaining and maintaining a patent, the right of protection for a utility model, the right from registration of an industrial design, incurred, among others, for translation, periodic fees, conduct of proceedings by the Patent Office;
7. Depreciation write-offs made in a given tax year on tangible and intangible assets used in research and development activities, excluding passenger cars and structures, buildings and premises under separate ownership;
8. Depreciation write-offs on intangible assets in the form of development costs (successfully completed).

If a taxpayer possesses the status of a research and development centre, then it can additionally apply depreciation deductions to structures, buildings and premises that are separately owned and used in business activities, as well as costs related to obtaining expert opinions, opinions, consulting services and contract research, technical knowledge and patents or a license for a protected invention. A taxpayer may deduct up to a maximum of 200% of eligible costs.

Conceptual work aimed at reducing the share of conventionally generated electricity consumption and shifting to the use of alternative energy sources (DNTI Interpretation of July 27, 2023) may be recognized as a research and development activity. Innovative work aimed at developing a storage system for self-generated electricity by the entrepreneur also meets the definition of R&D.

Likewise, the development of new and improved processes for order picking that are intended to reduce costs, as well as ecologically relevant, may constitute a research and development activity (DNTI interpretation of February 1, 2023).

Another area that has come under the spotlight for socially responsible activities is the environmental area. Among the countries that have introduced tax incentives for such activities are: (Review of Tax Incentives… 2011).

1. France introduced tax credits for R&D activities related to investment or implementation of environmental innovations of 30% of expenses incurred by existing companies and 35–40% for new companies.
2. Spain introduced a tax credit of 8% of the investment value for the purchase of technological innovations. The relief applies to, among other things, equipment for reducing air and noise pollution, equipment for reducing water pollution and that for reducing or disposing of industrial residues.
3. Belgium introduced a tax credit of 80% of the expenses invested in acquiring a patent, subject to certain conditions. The relief is available to all entities regardless of their legal form.
4. Greece favours production facilities that incur expenses to reduce the environmental impact of their operations. Such companies are authorized to reduce the tax base by 50% of the value of such expenses.
5. The United Kingdom has introduced increased tax depreciation rates for investments in selected "green" technologies, these are mainly energy-efficient devices and installations.

An example of the EU's support for environmental protection measures are subsidies and concessions for zero-emission industries. As of March 9, 2023, special rules were introduced to support entrepreneurs who invest in sectors dedicated to a carbon-free economy. They give member countries the opportunity to support domestic companies that fit into the EU's decarbonization policy and can accelerate the energy transition (http://psew.pl/dotacje-i-ulgi-podatkowe-dla-polskiego-przemyslu-bezemisyjnego-unia-daje-zielone-swiatlo/).

Another example of a relief for socially responsible activities is **the relief for sponsorship**. It provides an additional deduction of 50% of deductible expenses incurred for the following activities:

1. Sports,
2. Cultural activities within the meaning of the Act of October 25, 1991 on organizing and conducting cultural activities, and
3. Supporting higher education and science.

Among the deductible expenses are those incurred on:

1. Sports activities, i.e. costs incurred for financing:
   - A sports club,
   - A sports scholarship, and
   - A sports event that is not a mass sports event;
2. Cultural activities, i.e. costs incurred to finance:
   - Cultural institutions,
   - Cultural activities carried out by art universities and public art schools;
3. Activities supporting higher education and science, i.e. costs incurred for:
   - Scholarships,
   - Financing of fees to employees employed by the taxpayer,
   - Financing of the salaries, including derivatives, of students undergoing internships and apprenticeships at the taxpayer's company as required by the study curriculum,
   - Financing of dual studies, including the costs of internships,
   - Financing of dual studies, including the costs of internships,
   - Remuneration paid within 6 months from the date of employment by a taxpayer organizing apprenticeships for students of a given university to an employee who is a graduate student at this university employed through an academic career office.

Sponsorship is nothing more than a PR activity intended to create a good association with a brand as well as positively influence its image. Among the reasons for donating funds through sponsorship are (Grzelonska, 2016):

1. Business attitude: the goal is to make a profit, which can be achieved through access to a larger group of potential contractors and a new group of consumers;
2. Ethical attitude: the corporation wants merely to be associated as socially responsible;
3. Political attitude: by supporting specific cultural activities, the company wants to establish contacts with a group of people responsible for creating socio-political norms;

4. Stakeholder attitude: the corporation wants to positively influence the reputation and competitiveness of a particular region, which is expected to translate directly into a greater interest of possible employees.

The beneficiaries of the sponsorship relief are entrepreneurs earning income from non-agricultural business activities, the income from which is taxed on a general or flat tax basis, as well as the CIT taxpayers, earning income other than income from capital gains. The relief is therefore aimed at different groups, such as sole proprietors and partnerships.

Among the state incentives intended to encourage entrepreneurs to engage in socially responsible activities is the possibility of deducting from the income a donation made to a public benefit organization (PBO), with the value of the deduction not to exceed 6% of the annual income (PIT) and 10% (CIT).

The deductible donations are those made for, among other things:

1. Religious purposes,
2. Blood donation carried out by honorary blood donors,
3. Vocational training to public schools,
4. Public benefit purposes, including but not limited to (personal income tax law, art. 26 paragraph 1, item 9):
   - Social assistance, support for the family and the foster care system,
   - The provision of free legal aid,
   - Activities for the integration and professional and social reintegration of people at risk of social exclusion,
   - Charity activities,
   - Maintenance and dissemination of national traditions,
   - Activities for national and ethnic minorities and regional language,
   - Activities for the integration of foreigners,
   - Protection and promotion of health, including therapeutic activities,
   - Activities for the benefit of the disabled,
   - Promotion of employment and vocational activation of the unemployed and those at risk of redundancy,

- Activities for the equal rights of women and men,
- Activities for people of the retirement age,
- Activities supporting economic development, including the development of entrepreneurship.

Nowadays, socially responsible companies often support various types of institutions and associations engaged in charitable activities, which have the status of PBOs. The aforementioned tool is designed to encourage taxpayers to support charitable organizations. By making a donation, in fact, a company can reduce its income and revenue and thus pay less tax.

The tax concessions offered by the state under CSR can also be seen within the employee area. In Poland, an entrepreneur who employs an unemployed person over 50 years of age, is exempt from paying the Labor Fund and the Guaranteed Employee Benefits Fund on them for the first 12 months. There is also a possibility of obtaining various types of refunds.

In Belgium, there is a flat tax exemption for the employment of workers in managerial positions in the export and internal quality control departments. Companies can also count on income tax exemptions for hiring additional employees.

Many countries also offer preferences for employee training, allowing training expenses to be counted as tax deductible. Such opportunities are provided by the Polish, Norwegian or French legislation. In Belgium, on the other hand, it is possible to obtain subsidies for employee training.

5.4 Conclusion

National governments can make an important contribution to policy-making on sustainable development and social responsibility. CSR issues appear in various EU documents not directly dedicated to the subject. The EU's 2001 Sustainable Development Strategy, revised in 2006, states that public policy is crucial in promoting CSR. At the same time, all listed companies with at least 500 employees are encouraged to provide information on their economic, environmental and social performance in their annual reports (Communication from the Commission, 2001).

The state is one of the main entities able to compel companies to fulfil their social responsibilities. The state's main role in this regard is to create favourable and conducive conditions for the implementation of this idea by business entities.

With the beginning of the 1990s, Poland's development toward a market economy became noticeable. However, in terms of CSR, our country lagged far behind other EU countries. It is only in recent years that changes in this area have been noticeable. New tax credits have emerged to encourage companies to engage in socially responsible activities.

Tax credits are instruments that have a direct effect on increasing the profitability of investments in socially responsible activities, so they significantly affect the willingness of entrepreneurs to engage in CSR.

Summarizing the conducted analysis, it should be noted that in selected European countries there is a number of initiatives supporting CSR activities. Nevertheless, there are still areas that could be of interest to both entrepreneurs as well as government institutions.

References

Aarts, D. (2011). Społeczna odpowiedzialność przedsiębiorstw, w: Sztuka public relations. Z doświadczeń polskich praktyków, Związek Firm Public Relations, Warsaw.

Babiak, K. and Trendafilova, S. (2011). CSR i odpowiedzialność środowiskowa: Motywy i naciski na przyjęcie zielonych praktyk zarządzania, Społeczna odpowiedzialność biznesu i zarządzanie środowiskowe, 18(1), pp. 11–24.

Balzarova, M. A. and Castka, P. (2012). Wpływ interesariuszy i wkład w rozwój standardów społecznych: przypadek podejścia wielu interesariuszy do rozwoju ISO 26000, Journal of Business Ethics, 111, pp. 265–279.

Bereś, A. (2017). Budowanie wizerunku biznesu społecznie odpowiedzialnego przez pryzmat realizacji projektów, Studia Ekonomiczne, 332, pp. 110–121.

Bernatt, M. (2009). Społeczna odpowiedzialność biznesu. Wymiar konstytucyjny i międzynarodowy, Wydawnictwo Naukowe Wydziału Zarządzania Uniwersytetu Warszawskiego, Warsaw.

Bowen, H. R. (1953). Social Responsibilities of the Businessman, Harper and Brothers, New York.

Cansino, José M., et al. (2011) Promoting renewable energy sources for heating and cooling in EU-27 countries, Energy Policy, 39(6), pp. 3803–3812.

Carroll, A. B. (1991). The pyramid of Corporate Social Responsibility: Toward the moral management of organizational stakeholders, Business Horizons, 34(4), pp. 39–48.

Carroll, A. B. (2015). Corporate social responsibility, Organizational Dynamics, 44(2), pp. 87–96.

Communication from the Commission. (2001). A Sustainable Europe for a Better World: A European Union Strategy for Sustainable Development, Commission of the European Communities, Brussels, 15.05.2001, COM, 264, p. 9.

Dodd, E. M. (1932). For whom are corporate managers trustees?, Harvard Law Review, 45, p. 1145.

Duff, D. G. (2009). Unikanie podatków w XXI wieku, Thomsona.

Fujarska M. and Szewczyk P. (2011). Ocena metod integracji wybranych systemów zarządzania oraz rola odpowiedzialności społecznej, in Zeszyty naukowe Politechniki Śląskiej, s. Organizacja i Zarządzanie, v. 59, pp. 161–172, Wydawnictwo Politechniki Śląskiej, Gliwice.

Green Paper. (2001). European Commission: Directorate-General for Employment, Social Affairs and Inclusion, Promoting a European framework for Corporate Social Responsibility, Commission of the European Communities, Brussels, 18th July 2001.

Grzelońska, U. (2016). Ekonomiczna strona kultury. Instytut Nauk Ekonomicznych Polskiej Akademii Nauk, Warsaw.

Gołąbeska, E. and Harasimowicz, A. (2023). Wybrane problemy związane z realizacją systemów wykorzystujących zieloną energię, Oficyna Wydawnicza Politechniki Białostockiej, 143, DOI: 10.24427/978-83-67185-59-2.

Gomułowicz, A. and Małecki, J. (2013). Podatki i prawo podatkowe, LexisNexis Polska, Warsaw.

Jastrzębska, E. (2011). Państwo a społeczna odpowiedzialność biznesu, Kwartalnik Kolegium Ekonomiczno-Społecznego. Studia i Prace, 2, pp. 14–45.

Jha, A. and Cox, J. (2015). Corporate social responsibility and social capital, Journal of Banking & Finance, 60, pp. 252–270.

Kacem, H. and Omri, M.A.B. (2022). Corporate social responsibility (CSR) and tax incentives: the case of Tunisian companies, Journal of Financial Reporting and Accounting, 20(3/4), pp. 639–666. doi: 10.1108/JFRA-07-2020-0213

Leoński, W. (2016). Rola państwa i instytucji rządowych w promowaniu koncepcji społecznej odpowiedzialności biznesu w polsce, Research Papers of the Wroclaw University of Economics/Prace Naukowe Uniwersytetu Ekonomicznego we Wroclawiu, 449.

Mekhilef, S., Saidur, R. and Safari, A. (2011). A review on solar energy use in industries. Renewable and Sustainable Energy Reviews, 15(4), pp. 1777–1790.

Pop, O. and Catalin Martin, G. C. (2011). Promowanie społecznej odpowiedzialności biznesu na rzecz zielonej gospodarki i innowacyjnych miejsc pracy, Procedia-Social and Behavioral Sciences, 15, pp. 1029–1023.

Porter, M. and Kramer M. (2011). Creating shared value, Harvard Business Review, 89, nr 1–2 (styczeń–luty 2011): 62–77.

Rojek-Nowosielska, M. (2017). Definicja CSR według normy ISO 26000 a praktyka gospodarcza, Ruch prawniczy, ekonomiczny i socjologiczny, 79(3).
Shazmin, S. A. A., Sipan, I. and Sapri, M. (2016). Property tax assessment incentives for green building: A review, Renewable and Sustainable Energy Reviews, 60, pp. 536–548.
Sinenko, O. A. (2016). Methods of assessing of tax incentives effectiveness in special economic zones: an analytical overview, Journal of Tax Reform, 2(3), pp. 168–178.
Stephenson, D.and Vracheva, V. (2016). Corporate Social Responsibility and tax avoidance: A literature review and directions for future research, http://dx.doi.org/10.2139/ssrn.2756640 (Accessed: 12.03.2024).
Wasilewski, M.and Bischoff A. (2017). Unikanie Opodatkowania Przez Przedsiębiorstwa Międzynarodowe W Kontekście Społecznej Odpowiedzialności Biznesu (Multinational Enterprise Tax Avoidance within the Context of Corporate Social Responsibility). Available at SSRN 2957157, 187–198.

Internet sources:

https://www.cire.pl/pliki/2/12022Raport.pdf (Accessed: 25.01.2024).
https://www.gov.pl/web/fundusze-regiony/o-zespole (Accessed: 25.03.2024).
http://psew.pl/dotacje-i-ulgi-podatkowe-dla-polskiego-przemyslu-bezemisyjnego-unia-daje-zielone-swiatlo/(Accessed: 26.03.24).

Legal acts:

Corporate Income Tax Act of February 15, 1992 (Journal of Laws 1992, item 865).
Individual interpretation of the Director of National Tax Information dated July 27, 2023 (Ref. 0114-KDIP2-1.4010.235.2023.6.MR).
Individual interpretation of the Director of National Tax Information dated February 1, 2023 (Ref. 0114-KDIP2-1.4010.211.2022.4.MW).
Tax Ordinance Act of August 29, 1997 (Journal of Laws 1997, no. 137, item 926).

6

Green Solutions in the Smart City

RENATA WŁODARCZYK,
PAULINA KALEJA, JANUSZ STĘPNIK
AND BEATA BARSZCZOWSKA

6.1 Introduction

At present, the demand for electricity and heat is constantly growing in Poland and all around the world. The continuous urbanization of cities and agglomerations is causing increasing emissions of pollutants into the atmosphere. Due to the above, a very important issue recently is the search for a solution enabling the reduction of emissions and immissions of pollutants and the decarbonization of the most important industries. The high level of exploitation of natural deposits, which are currently used on a large scale, forces us to look for new zero-emission energy carriers. Therefore, actions should be taken that will have a positive impact on meeting the needs of consumers, i.e. electricity, heat, cooling, combined with environmental control.

This chapter presents an analysis of the possibility of using hydrogen as a zero-emission energy carrier and fuel used to power vehicles. The so-called hydrogen value chain comprises production, storage, transmission and use of hydrogen in the energy and transport industries in the smart cities of the future.

6.2 Hydrogen and Its Physico-chemical Properties as an Energy Carrier

Hydrogen is an element that exists in the form of a diatomic molecule (H_2). Hydrogen is a colorless and odorless gas. It is characterized by high calorific value, heat of combustion and high octane number, which constitutes its high energy potential. The heating value of hydrogen is three times higher than the heating value of gasoline.

DOI: 10.1201/9781003465157-6

In addition, it burns with a high flame, quickly and into water. Hydrogen is a flammable gas, and with oxygen in specific mass ratios it forms an explosive mixture (Pawlik and Strzelczyk, 2016). From the perspective of using hydrogen as an energy carrier, the key thing is that the product of hydrogen combustion is water, which is non-toxic to the environment (Chmielniak, 2020). Under normal conditions, hydrogen is a colorless, odorless and tasteless gas (Lewandowski, 2007). Depending on the temperature, H_2 may occur in three states of matter (Barycka and Skudlarski, 1981). Regardless of the form in which it occurs, hydrogen is the lightest chemical element. The density of H_2 in the gas phase is almost a thousand times lower than its density in the solid phase, while the specific heat in the solid state is 5.5 times lower than the gas state (Jastrzębska, 2007).

6.3 Hydrogen Economy: Hydrogen Value Chain

The hydrogen economy defined in the document of the Polish Hydrogen Strategy until 2030 with a perspective until 2040 is a broadly understood value chain relating to four issues and technical solutions, i.e. (PHS, 2021):

1. Production,
2. Transport,
3. Storage, and
4. Use.

The above logistic chain is of great importance in the development of hydrogen infrastructure. Among these four terms, none is more or less important in this process, because all points must be met to talk about fully preserved logistics necessary for the development of hydrogen technologies. The idea of using hydrogen as an energy carrier using renewable energy sources is presented in Figure 6.1 (Włodarczyk and Kacprzak, 2019). The possibility of using a fuel cell is determined by the source from which fuel for the generator can be obtained, and then by the method of storing the fuel. The ignition initiation energy of hydrogen (which makes combustion more efficient) and its reserves in the form of water are practically inexhaustible.

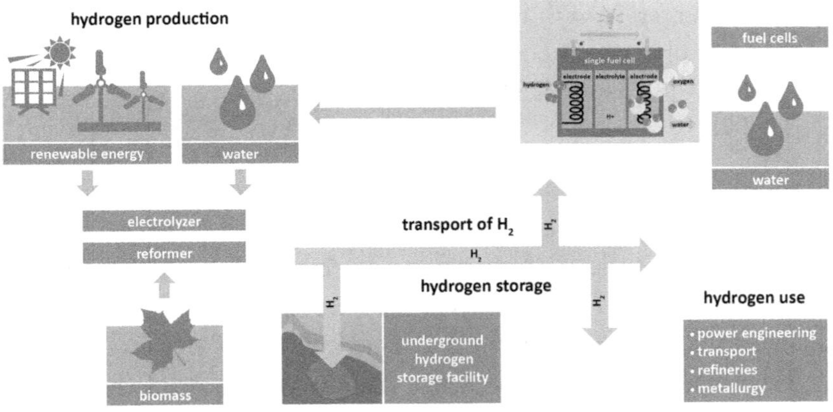

Figure 6.1 The idea of using hydrogen. (Own elaboration.)

6.3.1 Production of Hydrogen

The greatest potential for hydrogen production in Poland involves the use of fossil fuels, mainly hard coal and lignite. The estimated hydrogen production potential in Poland indicates that 45% of H_2 will be produced from hard coal, 50% from brown coal, 4% from biomass and natural gas, and 1% by other, probably expensive methods using water as a raw material (100% = 37 million tonnes/year) (Bartosik et al., 2016). Hydrogen can be obtained as a by-product in chemical reactions, which allows for the so-called by-product to obtain a new product, in this case hydrogen. The most common classification of hydrogen production processes includes the following methods (Besancon et al., 2009; Kothari, Buddhi and Sawhney, 2008):

1. Electrochemical,
2. Thermal,
3. Biological.

However, the detailed processes from which hydrogen is produced include (El-Shafi, Kambara and Hayakawa, 2019; Rozendal et al., 2006):

1. Steam reforming,
2. Thermal decomposition of water,
3. Electrolysis,
4. Gasification,
5. Pyrolysis,

6. Fermentation with microorganisms,
7. Anaerobic decomposition, and
8. Photoelectrochemical decomposition.

All activities of energy and transport companies, ventures and other activities aimed at the development of hydrogen technologies in Poland should now be paid attention to how all hydrogen tycoons classify themselves in terms of the structure of their share in the hydrogen market. Figure 6.2 shows a chart showing the structure of share in the hydrogen market in Poland in 2020.

From the chart above, we can conclude that most of the precursors in hydrogen production are primarily chemical and fuel and refinery companies. Of course, it should be taken into account that hydrogen in these plants is mostly produced as a by-product. For now, grey hydrogen is mainly produced in Poland. However, this market is growing at a galloping pace. Large financial outlays from the European Union (EU) allocated to the development of hydrogen technologies significantly facilitate research and new inventions for the development of this technology. The share of raw materials involved in hydrogen production is quite diverse, almost half of H_2 is obtained from natural gas, and to a lesser extent from crude oil and coal (Table 6.1). The use of electrolysis seems to make sense only if the electricity necessary for the reaction comes from renewable energy sources (https://h2poland.eu, 2023). Figure 6.3 shows the

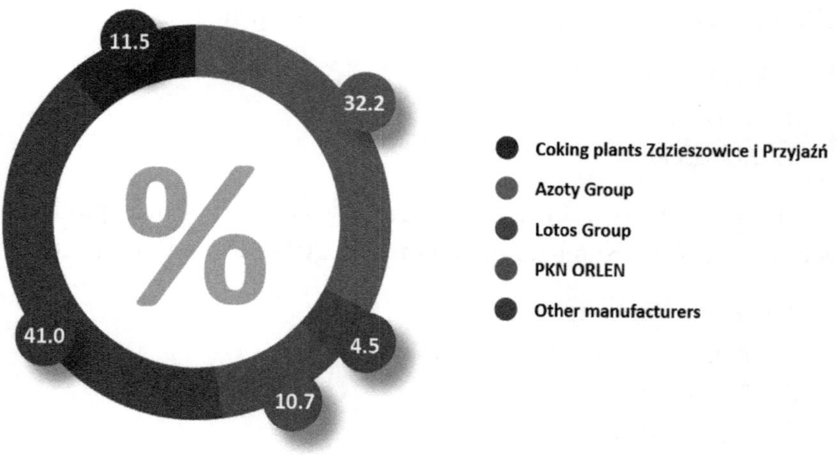

Figure 6.2 Structure of share in the hydrogen market in Poland in 2020. (Own elaboration.)

Table 6.1 Characteristics of the Technology for Producing Hydrogen from Fossil Fuels and Biomass

	PROCESS	FUEL	TOTAL EFFICIENCY [%]	COST OF H_2 [USD/GJ]	INVESTMENT COSTS [USD/GJ H_2]
Steam reforming	• Well known • Extensive infrastructure	Natural gas	65–751	5–8	9–15
Partial oxidation	• Cheap fuel	petroleum	501	7–10	9–22
Gasification	• Poor infrastructure	Coal	42.5–46.52	10–12	33–34
Pyrolysis	• Simultaneous CO_2 separation	Biomass		9–13	20–42
	• Concentrated CO_2 stream ideal for sequestration process		47.92	9–13	15–19

Source: Based on Chmielniak T, Hydrogen energy, 1st edition, Wydawnictwo Naukowe PWN, Warszawa 2020.

ranking of seven EU countries producing hydrogen, given in the unit [million tonnes per year].

Further development of hydrogen production technology will undoubtedly be related to obtaining it using renewable sources, especially water electrolysis. Currently, this is not a competitive method and is closely related to gas, electricity prices and CO_2 emission fees, which requires additional consistency in climate policy (https://h2poland.eu, 2023).

Polish companies have developed individual development strategies in which the use of hydrogen is an integral part. Poland is currently

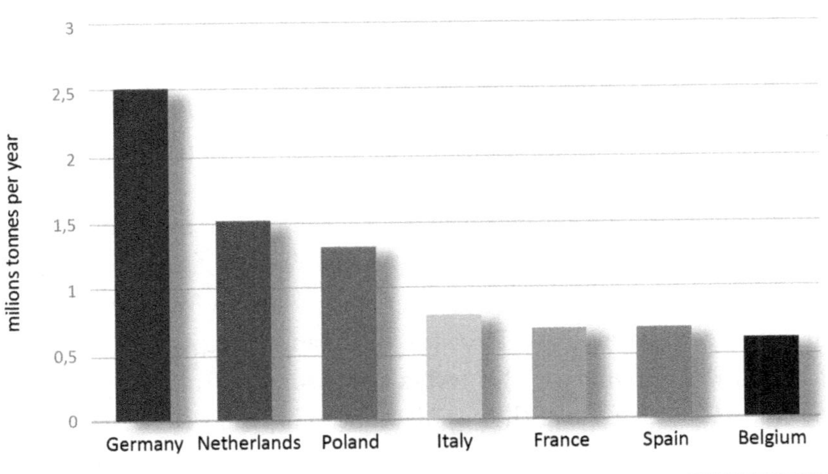

Figure 6.3 Hydrogen production in Europe (end of 2018). (Own elaboration.)

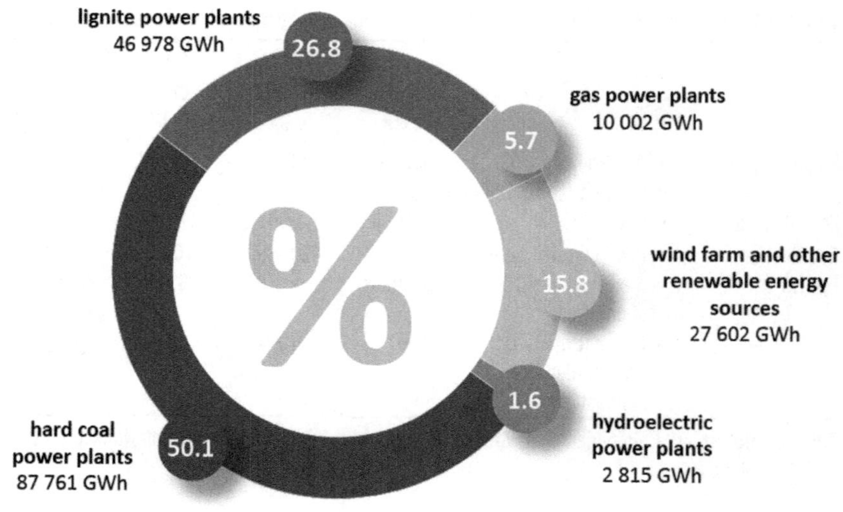

Figure 6.4 Structure of electricity production in Poland in 2022. (Own elaboration.)

the fifth largest producer of hydrogen in the world and the third in the EU (1.3 million tonnes per year), but the main problem is the method of its production, which is more than 95% from fossil fuels (Demusiak, 2012). Figure 6.4 presents data on electricity production from conventional sources and renewable energy sources. Renewable sources produced 27,602 GWh, which constituted 15.8% of the total national production for 2022. Taking into account the data contained in the Polish Hydrogen Strategy that 50 kWh of electricity is needed to produce 1 kg of hydrogen, it can be easily calculated that to produce 1.3 million tonnes of H_2 require 65 TWh, so the entire power from renewable energy sources would not be enough to achieve even half of this task. The total annual electricity production in Poland in 2022 amounted to 175,157 GWh (www.rynekelektryczny.pl).

According to the government strategy and the 2021 Green Hydrogen from Renewable Energy Report prepared by the Lower Silesian Institute for Energy Studies, offshore wind farms and photovoltaics supplementing them are to be the main flywheel for powering the electrolysers. As coal-based mines and energy are phased out, nuclear power plants are also to take over some of their responsibilities (PEP, 2040). One of the popular methods using renewable energy sources is electrolysis. This is a process in which water is separated directly into hydrogen and oxygen molecules under the influence of current supplied from an external power source. The carriers of electric charges in electrolysis are anions and cations. This process is carried out using a device called an electrolyser. When obtaining hydrogen using the classic electrolysis method, more energy is used than could be obtained later from hydrogen, because 3 kWh of electricity is needed to obtain 1 m^3 of hydrogen (www.gov.pl, 2024). However, it should be emphasized that 100 million m^3 of "hydrogen" fuel corresponds to 25,000 tonnes of crude oil.

6.3.2 Transportation of Hydrogen

The International Energy Agency considers the most optimal method of transporting hydrogen (up to 1,500 km) by the use of dedicated pipelines, which are already used in Western European countries (https://h2poland.eu/pl/, 2023). On intercontinental routes on ships in the form of e.g. ammonia, hydrogen is most often transported using car sets with semi-trailers - equipped with containers with bundles of cylinders, pipe tanks or tanks for transporting liquid cryogenic hydrogen. The pipeline trucks are capable of loading 300–500 kg of compressed hydrogen gas (pressure 200–250 bar). Modern cylinder trucks enable loading up to 900 kg of compressed hydrogen gas (pressure 500 bar), and tankers up to 3,500 kg of liquid hydrogen (https://h2poland.eu, 2023).

The use of hydrogen in transport and hydrogen transport is another extremely important element in the entire logistics structure. Although this is one of the most difficult points to meet in the hydrogen economy. Currently, when it comes to transporting hydrogen, what we know is that this gas cannot be transported alone, i.e. pure, but only in admixtures due to its corrosive properties. This is an alternative action that aims to incorporate other molecules that are easier to transport. Such chemical compounds include ammonia (NH_3) and liquid organic hydrogen carriers (LOHC) (Report, 2021).

Road transport can be used to transport hydrogen, namely tankers, tubular tanks or tankers for the transport of liquid hydrogen. All transport must be based on the ADR Agreement, i.e. the agreement on the international carriage of dangerous goods by road. Of course, loading and transport must take place in accordance with strictly defined conditions. Another idea is to transport hydrogen by sea, i.e. the so-called IMDG.

6.3.3 Storage of Hydrogen

Hydrogen storage is another challenge in the world of technology and science. Hydrogen storage can be divided into physical and material. The undeniable advantage of hydrogen compared to electricity is the possibility of storing it for a longer period. As mentioned earlier, this element can be stored in solid, liquid or gaseous form. Due to transport and storage problems, most hydrogen-using installations are immediately connected to H_2-producing installations. As small hydrogen molecules tend to penetrate materials, tanks are required to create an effective barrier resistant to mechanical damage, such as high-density steel.

The most effective method of storing hydrogen is high compression, because available tanks ensure minimal loss of stored gas. This method is effective and well known in passenger cars and buses. The pressure value depends on the application and in the first case the tank maintains a pressure of 70 MPa, i.e. 700 bar, while in the case of buses and trains it is 35 MPa, i.e. 350 bar (https://h2poland.eu, 2023). For example, three-layer tanks of the popular Toyota Mirai car, three in number, are made mainly of carbon fibre with a capacity of 5.6 kg each. Light and very durable tanks do not pose a threat to passengers or the vehicle if they leak, because light hydrogen quickly escapes upwards. A significant disadvantage of storing hydrogen in gaseous form is the fact that such a tank would have to be of a large size, which is not economically justified (www.toyota.pl, 2023).

Maintaining hydrogen in a liquid state requires cooling to –253°C and the use of multi-layer cryogenic tanks, containing a vacuum layer to ensure thermal insulation and safety valves. Maintaining such temperature and storage conditions requires large energy inputs of 25–35% necessary for cooling (https://h2poland.eu/pl/, 2023). By creating a chemical compound with a metal or alloy, the so-called metal hydride, hydrogen can be stored as a solid by coming into contact with the surface of the storage material, hydrogen molecules decay into atomic hydrogen and penetrate the material. However, the disadvantage of this solution is the relatively large mass of the storage material compared to the absorbed hydrogen (www.tuv.com, 2023).

Polish Hydrogen Strategy (PSW) until 2030 with a perspective until 2040. indicates that the most economical method will be underground storage facilities such as depleted oil and gas fields, abandoned mines and salt caverns (Tarkowski, 2017). Although exploited oil and gas deposits are larger, their porous geological structure means that the medium stored in these reservoirs would have to be additionally cleaned before their final use, which further increases costs and complicates the readability of the system. According to the 2021 Report Green hydrogen from renewable energy in Poland (Report, 2021), in Polish conditions it seems the most convenient to use depleted salt mines for technical reasons (high flexibility of work, possibility of long-term storage of H_2), safety (low porosity, does not contain water, chemically neutral in relation to the stored gas) and, in addition, the lowest storage costs have been proven. The same report, citing other studies, determined Poland's current potential for storing hydrogen in caverns at 2.2 TWh, while these needs will increase significantly in the following years, and so by 2030 it will be 3.6 TWh and approximately 36.5 TWh by 2050 (the last two figures apply to all types of warehouses) (Report, 2021).

To summarize, hydrogen storage methods include (www.deloitte.com, 2023):

1. above-ground pressure tanks,
2. storage in liquid form,
3. use of underground caverns and depleted fossil fuel deposits,
4. chemical storage methods.

Hydrogen tanks are subject to the Transport Technical Supervision of the TDT and when used in the automotive industry as a power source, they are subject to periodic and ad hoc tests. A tightness test and external inspection should be carried out once a year. However, when it comes to internal inspection and pressure test once every 10 years. All requirements in this matter are included in the Regulation of the Minister of Transport of October 20, 2006 on the technical conditions for technical supervision in the design, manufacture, operation, repair and modernization of specialized pressure equipment (Journal of Laws of 2014, item 1465) (www.poland.eu, 2023). Table 6.2 presents hydrogen storage methods, including processes and chemical compounds.

The next milestone is the invention of the first innovative hydrogen tanks in Europe from a Polish manufacturer, namely Sunex.

Table 6.2 Hydrogen Storage Methods

HOW CAN HYDROGEN BE STORED	
PHYSICAL METHODS	MATERIAL METHODS
• Compression, • Chilled gas, • Liquid hydrogen.	• Adsorption, • Organic compounds, • Interstitial hydrides, • Hydrides, • Chemical compounds.

Source: www.cire.pl, 2024.

This company intends to produce composite tanks from carbon fibres. This is another example of us trying to introduce green innovations. The next alternative is to store hydrogen in salt caverns. As it turns out, salt caverns have great potential as a form of storage (Tarkowski, 2017; Włodarczyk and Kacprzak, 2019). Salt caverns, which are a potentially attractive place for storing hydrogen, belong to the pore-fissure type. These are artificial chambers created by leaching salt from seam deposits or in salt domes. These types of chambers are used to store natural gas and petroleum products. Salt walls are impermeable and provide good gas insulation, the plastic properties of salt protect caverns against the appearance and spread of cracks, thus eliminating the risk of loss of tightness. Storing hydrogen in salt caverns allows this medium to be stored for up to several weeks. Depleted oil and gas deposits can also be used to store hydrogen. The level intended for its storage must have appropriate porosity and permeability, and the overburden rocks must ensure tightness against gas leakage to the surface. The principle of storing hydrogen in deep aquifers is based on similar assumptions as in the case of depleted oil and gas deposits, with the difference that the rock matrix contains brine instead of hydrocarbons. Polish places where hydrogen storage is possible are Mogilno (capacity of 585.4 million m^3 of natural gas) and Kosakowo (239.4 million m^3 of natural gas) are shown in Figure 6.5.

6.3.4 Hydrogen Use

Hydrogen is used in many industries and even in our everyday lives. We increasingly see passenger cars with engines based on fuel cells and electric buses powered by hydrogen on our roads (Melis, 2002).

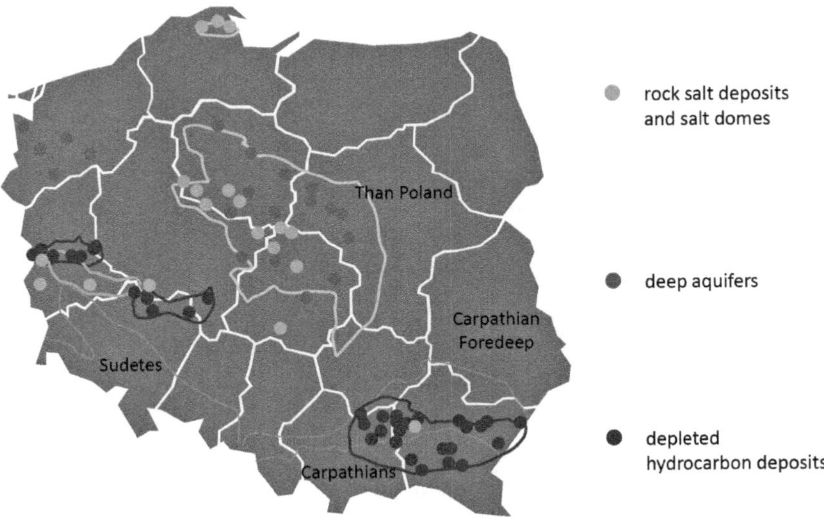

Figure 6.5 Location of potential hydrogen storage areas in rock salt deposits and salt domes as well as in depleted crude oil and natural gas deposits. (Own elaboration.)

Taking into account the increasing popularity of brands introduced to the market, the increase in hydrogen charging stations, this trend should only increase in the coming years. Hydrogen is used in the oil industry in such reactions as reforming with the intention of increasing the octane number; hydro-refining, which reduces the content of harmful compounds, thus stabilizing the fuel; or hydro-cracking, converting heavier oil fractions such as greases into gasoline and light oils. In the energy industry, in the power range of 200–1200 MW; direct hydrogen cooling of iron and rotor windings of domestic generators TGHW-360 and TWW-200 of conventional power plants has been used and is still used. In nuclear power plants, deuterium, one of its isotopes, serves as an electron retarder called a moderator. The agricultural industry uses it to produce ammonia, used as an artificial fertilizer, and in the food industry to hydrogenate simple vegetable oils, which are healthier than those of animal origin in products such as margarine or salad dressings. In jewellery, hydrogen-oxygen torches are used for precise welding and cutting, enabling the processing of gold, silver and platinum. During smelting and alloying in steel production, the use of H_2 instead of coal allows for the reduction of carbon dioxide emissions in the metallurgical industry.

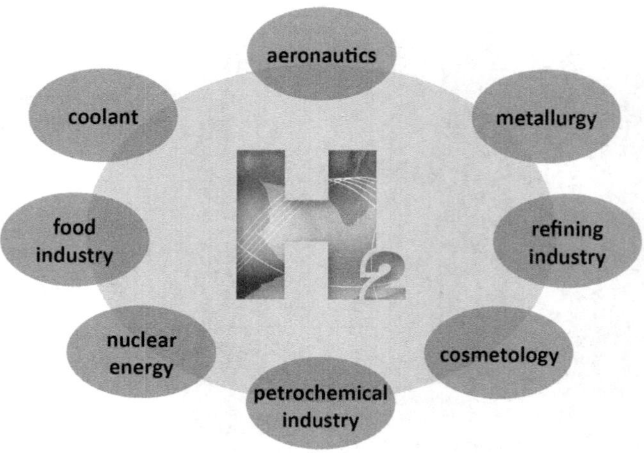

Figure 6.6 Hydrogen and its use in various economic sectors. (Own elaboration.)

The examples presented clearly show that hydrogen has been with us for a long time, although we may not have been fully aware of it. Considering that as an energy carrier it is completely neutral to the environment, its potential has not yet been fully exploited. If we also take into account EU directives, this gas, obtained in addition from renewable energy sources, is to play a more important role not only in transport but also in heating buildings.

Figure 6.6 shows hydrogen and its use in various economic sectors. As we can see, this gas is used in many areas necessary for the proper and everyday functioning of society.

6.4 Legal Aspects of the Implementation of Hydrogen Technologies

One of the most important goals of the state is to ensure the country's energy security, ensuring the competitiveness of the economy, energy efficiency and reducing the impact of the energy sector on the environment, while optimally using its own energy resources. The strategic document presenting the government's long-term strategy for the development of the energy sector is the Polish Energy Policy until 2040. PEP 2040 resulting from the Strategy for Responsible Development (SOR) is consistent with the National Energy and Climate Plan for 2021–2030. It is based on three pillars with eight specific objectives along with actions and strategic projects to achieve them. The implementation of the goals will allow for low-emission energy transformation to be carried out in an innovative, socially acceptable way, with respect for the environment.

The Polish Hydrogen Strategy (PSW) until 2030 with a perspective until 2040 is a strategic document that defines the main goals of the development of the hydrogen economy in Poland and the directions of intervention that are desirable to achieve them (PSP-2021, 2021). It is part of global, European and national activities aimed at achieving a low-emission economy (KP-2023, A hydrogen strategy-2021). PSW is consistent with the SOR and is consistent with its objectives, as well as with the activities presented in the strategy entitled Poland's energy policy until 2040 (PEP, 2040). Among European countries, Poland is the leader in hydrogen production, but it comes mainly from steam reforming of hydrocarbons used for industrial processes.

Its main producers are large state-owned industrial plants such as Orlen, Grupa Azoty S.A. (Orlen, 2024). The development of the hydrogen economy will therefore require the creation of new renewable energy generation capacities, hydrogen production and storage installations, electrolysers, fuel cells, transmission and distribution infrastructure and refuelling stations. However, it should be remembered that the document provides for support for a low-emission hydrogen economy.

The strategy sets six key goals (PEP 2040):

1. Energy and heating;
2. Use of hydrogen in transport;
3. Decarbonization of the economy;
4. New technologies;
5. Transmission, storage and distribution;
6. Appropriate legal environment.

The following are the key indicators to be achieved for 2030:

1. Installed capacity of the installation for the production of low-emission hydrogen at the level of 2,000 MW;
2. Number of hydrogen valleys not less than 5, creation of the Hydrogen Technology Centre, number of hydrogen stations at 32;
3. At least 1,000 hydrogen buses.

Currently, there are 10 hydrogen valleys in Poland. Figure 6.7 shows a map of Poland, showing the distribution of hydrogen valleys in Poland. Stakeholders of individual hydrogen valleys in Poland

1. Lower Silesian Hydrogen Valley
2. Mazovian Hydrogen Valley
3. Silesia – Lesser Hydrogen Valley
4. Subcarpathian Hydrogen Valley
5. Central Hydrogen Cluster named after Laszczynski Brothers
6. Pomeranian Hydrogen Valley
7. Greater Poland Hydrogen Valley
8. West Pomeranian Hydrogen Valley
9. Amber Hydrogen Valley
10. Agricultural Hydrogen Valley

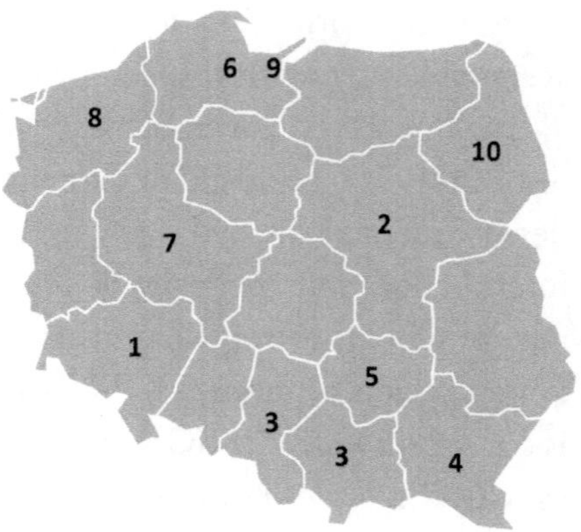

Figure 6.7 Distribution of hydrogen valleys in Poland. (Own elaboration.)

are constantly working on modern hydrogen technologies and their improvement. They obtain subsidies from the EU for this purpose, including development of research, modern technologies and adaptation of infrastructure in the country. In addition, some large companies that are stakeholders in hydrogen valleys are working intensively on adapting their installations and modernizing them in order to achieve the highest possible efficiency in hydrogen production or its recovery (Kaleja, 2023; Włodarczyk and Kaleja, 2023).

One of the companies that focuses on the development of hydrogen technologies is ORLEN S.A., which wants to use the potential of hydrogen fuel in four areas.

Figure 6.8 shows the road map and plans for the future for the development and use of hydrogen technologies, especially in broadly understood transport.

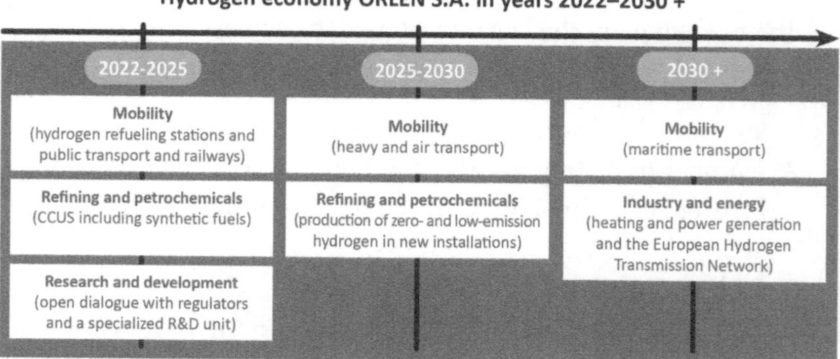

Figure 6.8 Hydrogen economy ORLEN S.A. in years 2022–2030 + in four areas of activity. (Own elaboration.)

6.5 Hydrogen Technologies in the Smart City Concept

In order to meet the requirements of the Paris Agreement of 2015 obliging to reduce greenhouse gas emissions, new solutions are necessary in terms of the development of zero and low-emission electricity sources. The implementation of this strategy does not solve the key issue of ensuring continuity of power supply from renewable energy sources dependent on weather conditions. Therefore, it seems to be the right solution to create appropriate conditions for the production and storage of hydrogen while creating appropriate logistic conditions. The implementation of industrial methods of water decomposition in an ecological way using electrolysers powered by renewable energy sources perfectly solves the problems of climate change (IEA, 2022). Green hydrogen is a "clean," inexhaustible carrier and its potential has been known for a long time. One of the main obstacles to its use was the high profitability of such projects. However, considering that in the past 20 years this ratio has dropped from 40 to 2 compared to the production of petroleum fuels, the prospects for achieving profitability have increased dramatically. Mass production of electrolysers will only accelerate this process, as will the rising prices of CO_2 emission allowances.

In line with the idea of using hydrogen technologies using electricity produced from renewable energy sources, we produce green – zero-emission hydrogen using water electrolysis. The electrochemical process splits water into hydrogen and oxygen. Hydrogen moves to the negatively charged pole and oxygen to the positively charged pole. This is currently the most promising method of producing ecological hydrogen. Renewable energy, while having many advantages, unfortunately has a significant drawback, which is instability resulting from weather conditions. During periods of increased renewable energy activity, the energy they produce becomes even problematic. Electricity that is not in demand affects the network voltage too high, its quality and also creates the risk of overload. The most rational solution here is electricity storage. Many of them, such as batteries and thermal heat storage units, fulfil their role for short periods, unlike hydrogen, which can be stored seasonally.

The rapidly rising costs of CO_2 emissions, falling prices of electrolysers, which have decreased by approximately 60% in the last decade, combined with the increase in their efficiency, as well as the prices of photovoltaic panels, which have decreased by approximately 80% since 2010, are the reasons this technology is becoming more and more interesting for business and public service providers. According to data from the Energy Institute – Research Institute, the uniform cost of producing hydrogen using alkaline electrolysers and on-grid photovoltaics is currently approximately USD 3–7.5/kg (www.europarl.europa.eu)

The use of hydrogen as an energy carrier in transport, which is burned flamelessly in fuel cells, is implemented by many car companies. Taking into account the price of refuelling a hydrogen car and the driving range, it is observed that hydrogen cars constitute serious competition for electric cars. Cars powered by hydrogen fuel cells, i.e. fuel cell electric vehicles (FCEVs) have been in the spotlight of the automotive industry several times, but this technology has always been considered a technology of the distant future. For this reason, although a lot of time has passed since the construction of the first cell-powered vehicle in 1966, these cars have not yet gained as much popularity as electric cars with batteries such as electric vehicles (EVs), battery electric vehicles (BEVs) or hybrid electric vehicles (HEVs). However, the largest car companies (Toyota, Hyundai, Honda, BMW, etc.) are still working on the development of fuel cell technology in cars. In some countries, such cars have been available directly from the manufacturer for many years.

A smart city is a city that uses advanced technologies to improve the functioning and quality of life of its inhabitants (Prawelska-Skrzypek and Blecharczyk, 2022). Zero-emission hydrogen as a fuel for public transport or even private transport will clearly contribute to reducing greenhouse gas emissions that are emitted in the process of burning non-renewable fuels. Investments in the development of transport and hydrogen infrastructure are a driving force for sustainable development and improving the quality of life, while caring for natural resources and at the same time caring for human health and natural resources.

The development of hydrogen technologies in the smart city concept may in the future constitute the basis for strategic planning in the city thanks to the support of innovation processes and citizen involvement in solving urban problems of its inhabitants. The global smart cities market based on the hydrogen value chain will implement the assumptions of an intelligent economy, fuel management, transport and intelligent energy with the support of information and communication technologies.

It should be emphasized here that preparing cities in the smart city concept requires taking a number of actions, not only infrastructural, but also legal and informational. The basis is the implementation of the hydrogen value chain, in particular the production of hydrogen, especially from renewable energy sources, and its storage. Transporting hydrogen from remote areas of the city will increase the costs of hydrogen refuelling, to the detriment of hydrogen vehicle users. Toyota Miray second generation consumes 0.84 kg of H_2/100 km, which gives us a price of approximately PLN 58/100 km. According to the data, the price of green fuel was PLN 69/kg of hydrogen, which indicates that the cost of full refuelling would be approximately PLN 385 (www.toyota.pl/swiat-toyoty/nowosci/2-generacja-toyoty-mirai) (the vehicle is equipped with three gas tanks that can hold 5.6 kg of H2).

We can therefore safely say that refuelling with green hydrogen is at an acceptable level, but one of the main reasons for the limited number of vehicles on European roads is the purchase price. However, taking into account the EU Energy Policy, falling cell prices and the competitiveness of producers, the price should strive for a rational level in the coming years. It should also be remembered that in Poland there are already the first hydrogen stations belonging to the Orlen group, but in this case the hydrogen fuel comes from fossil hydrocarbon sources. The multi-energy group itself plans to allocate over PLN 7 billion for this purpose by 2030 and build over 100 fast-charging stations in our part of Europe with hydrogen from renewable energy sources and municipal waste processing technology (https://www.orlen.pl/pl/o-firmie/o-spolce/dotacje/projekty-inwestycyjne/instrument-laczac-Europe-CEF/Clean-Cities-Hydrogen-Mobility-in-Poland-Phase-I/Clean-Cities--hydrogen-mobility-in-Poland-Phase-I-nr-projektu-INEACEFTRANM201923594741).

6.6 Conclusions

Hydrogen undoubtedly has a number of advantages that make it the fuel of the future. Taking into account environmental protection issues, its production using ecological technology fits perfectly into government and European strategies. The growing industrial demand for hydrogen and electricity in accordance with the Polish Energy Policy until 2040 (PEP, 2040) adopted by the Government of the

Republic of Poland and the Polish Hydrogen Strategy until 2030 with a perspective until 2040 (PSW) requires the use of renewable energy sources for the production of H_2 in a zero- and low-emission way, using surplus energy, mainly wind farms, and supporting them with dedicated PV farms. Ambitious government plans assume that the electrolyser capacity will reach 2–4 GW by the end of this decade, which will enable the implementation of the hydrogen value chain and expand the possibilities of its use, because the biggest challenges facing the development of green hydrogen are reducing the costs related to production, transport and storage, as well as the electrolysers themselves and the infrastructure cooperating with them. The idea of using hydrogen is based on closely related elements that create the so-called hydrogen economy: fuel production, fuel transport, storage and use of hydrogen in the economy for energy and/or transport purposes. As part of the International Partnership for Hydrogen and Fuel Cells in the Economy (IPHE) initiative, member states have the opportunity to exchange information between partners allowing for the implementation of effective and efficient international research and other activities related to the development of hydrogen and fuel cell technologies. The main goal of activities carried out under this initiative is to accelerate the development and implementation of hydrogen technologies, including enabling global energy security, as well as obtaining environmental and economic benefits related to the development of innovative technologies based on hydrogen and fuel cells.

In order to utilize hydrogen as a zero-emission fuel and implement smart city assumptions, it is necessary to solve the following issues:

1. Construction of appropriate infrastructure (network) of hydrogen refuelling stations,
2. Competitive price of hydrogen fuel,
3. Introduction of financial support for the purchase of an FCEV car with facilities for vehicle owners.

References

Bartosik, M., Kamrat, W., Kaźmierkowski, M., Lewandowski, W., Pawlik, M., Peryt, T., Skoczkowski, T., Strupczewski, A. and Szeląg, A. (2016). Electricity storage and hydrogen economy, Przegląd Elektrotechniczny, 92(12), pp. 332–340.

Barycka, I. and Skudlarski, K. (1981). Fundamentals of Chemistry, PWN, Warsaw.

Besancon, B. M., Hasanow, V., Imbault-Lastapis, R., Be-nesh, R., Barrio, M. and Moelnvile, M. J. (2009). Hydrogen quality from decarbonized fossil fuels to fuel cells, International Journal of Hydrogen Energy, 34(5), pp. 2350–2360.

Chmielniak, T. (2020). Hydrogen Energy, 1st edition, Wydawnictwo Naukowe PWN, Warsaw.

Demusiak, G. (2012). Obtaining Hydrogen Fuel by Reforming Natural Gas for Low-Power Fuel Cells, Nafta-Gaz, 10, pp. 671–672.

El-Shafi, M., Kambara, S. and Hayakawa Y. (2019). Hydrogen production technologies overview, Journal of Power and Energy Engineering, 7, pp.107–154.

Jastrzębska, G. (2007). Renewable Energy Sources and Pro-ecological Vehicles, WNT, Warsaw.

Kaleja, P. (2023). Analysis of the development of Polish hydrogen valleys in the context of the assumptions of the "Polish Hydrogen Strategy until 2030 with a perspective until 2040". Diploma thesis.

Kothari, R., Buddhi, D. and Sawhney, R. L. (2008). Comparison of environmental and economic aspects of various hydrogen production methods, Renewable and Sustainable Energy Reviews, 12, p. 553.

Lewandowski, W. M. (2007). Pro-ecological Renewable Energy Sources, WNT, Warsaw.

Melis, A. (2002). Green alga hydrogen production: Progress, challenges and prospects, International Journal of Hydrogen Energy, 27, pp. 1217–1228.

Pawlik, M. and Strzelczyk, F. (2016). Power Plants, WNT, Warsaw.

Prawelska-Skrzypek, G. and Blecharczyk, W. (2022). Smart and Sustainable Cities in Management Theory and Practice, IGSMiE PAN Publishing House, Kraków.

PSP-2021. (2021). Polish Hydrogen Strategy until 2030 with a perspective until 2040, Ministry of Climate and Environment, Warsaw.

Report. (2021). Green hydrogen from renewable energy sources in Poland. The use of wind energy and PV to produce green hydrogen as an opportunity to implement the assumptions of the EU Climate and Energy Policy in Poland. Wrocław.

Rozendal, R. A., Hamelers, H. V. M., Euverni, k G. J. W., Metz, S. J. and Buisman, C. J. N. (2006). Principle and perspective of hydrogen production through biocatalysed electrolysis, International Journal of Hydrogen Energy, 31, pp. 1632–1640.

Tarkowski, R. (2017). Perspectives of using the geological subsurface for hydrogen storage in Poland, International Journal of Hydrogen Energy, 42(1), pp. 347–355.

Włodarczyk, R. and Kacprzak, A. (2019). Storing energy in the form of hydrogen in Polish conditions: Potential and challenges, in Majchrzak-Kucęba I., Mirek P. and Bień J. (eds.), Modern Technologies of Energy Conversion and Storage. Częstochowa University of Technology Publishing, Częstochowa.

Włodarczyk, R. and Kaleja, P. (2023). Modern hydrogen technologies in the face of climate change-analysis of strategy and development in Polish conditions, Sustainability, 15, p. 12891. https://doi.org/10.3390/su151712891

Internet source:

A hydrogen strategy for a climate-neutral Europe, Communication from the Commission to the European Parliament, Brussels 07/08/2020, Texts adopted - European hydrogen strategy - Wednesday, May 19, 2021 (europa.eu) (Accessed: July 2023)

https://h2poland.eu/pl/kategorie/przesyl-magazynowanie/zbiorniki-na-wodor/zbiorniki-na-wodor/ (accessed 11/10/23)

https://www.cire.pl/artykuly/materialy-problemowe/wodor–zdrowie-zloto-bezpieczne-magazynowanie–szanse-i-wyzywania-cziv (Accessed: 18. 02. 2024)

https://www.orlen.pl/pl/o-firmie/media/komunikaty-prasowe/2022/luty/Grupa-ORLEN-inwestuje-w-wodor (Accessed: 01.12. 2023).

https://www.rynekelektryczny.pl/produkcja-energii-elektryczna-w-polsce/ (Accessed: March 2024)

https://www.toyota.pl/swiat-toyoty/nowosci/2-generacja-toyoty-mirai (Accessed: 09.12.2023)

https://www.tuv.com/landingpage/pl/hydrogen-technology/main-navigation/storage/ (Accessed: 13.12. 2023)

https://www2.deloitte.com/pl/pl/pages/tax/articles/strefa-ulg-i-dotacji/Technologie-wodorowe-Magazynowanie-wodoru.html (Accessed: 20. 11.2023)

https://wysokienapiecie.pl/32899-kto-zarobi-na-polskim-wodorze/ (Accessed: 07.05. 2023)

IEA. (2022). Global Hydrogen Review 2021, International Energy Agency, www.iea.org (Accessed: 01.09.2022)

KP-2023. National Energy and Climate Plan, https://www.gov.pl/web/klimat/national-energy-and-climate-plan-for-the-years-2021-2030 (Accessed: 08.07. 2023).

Orlen. (2024). ORLEN Group to launch international hydrogen program | PKN ORLEN (Accessed: March 2024)

Poland's Energy Policy until 2040 (PEP 2040), https://www.gov.pl/documents/33372/436746/PEP2040_projekt_v12_2018-11-23.pdf/ee3374f4-10c3-5ad8-1843-f58dae119936 (March 2024)

Polish Hydrogen Strategy until 2030 with a development perspective until 2040. Annex to Resolution No. 149 of the Council of Ministers of November 2, 2021 (item 1138). Warsaw, October 2021. Ministry of Climate and Environment

www.europarl.europa.eu https://www.europarl.europa.eu/Renewable hydrogen: what are the benefits for the EU? (Accessed: August 2023) www.gov.pl, 2024, LOTOS Green H2 project approved by the European Commission - Ministry of Climate and Environment - Gov.pl Portal (www.gov.pl)

www.toyota.pl, 2023 https://www.toyotapl.com/articles/2021/toyota-mirai-drugiej-generacji (Accessed: 10.12.2023)

For Product Safety Concerns and Information please contact our EU representative GPSR@taylorandfrancis.com Taylor & Francis Verlag GmbH, Kaufingerstraße 24, 80331 München, Germany

Batch number: 10397790

Printed by Printforce, the Netherlands